# Railroad Spikes

*A Collector's Guide*

JAMES M. JOYCE

SUTTER HOUSE

Copyright © 1985 by James M. Joyce

*All rights reserved*

No part of this book may be reproduced in any form
or by any electronic or mechanical means including
information storage and retrieval systems without permission
in writing from the publisher or author, except by a reviewer
who may quote brief passages in review.

LIBRARY OF CONGRESS CATALOGING IN PUBLICATION DATA

Joyce, James M., 1940-
 Railroad spikes.

 Bibliography: p.
 1. Nails and spikes—Collectors and collecting.
I. Title.
TS440.J69   1985         625.1'5         85-2805
ISBN-10: 091501033X

SUTTER HOUSE
P.O. Box 212
Lititz, Pa. 17543

PRINTED IN THE UNITED STATES OF AMERICA

*About the cover . . .*
 The cover art is a collage of railroad spike patents and advertisements from a design concept by Jean Joyce.

To my wife
Jean
and our daughters
Tannisse and Susan
and our son
Michael

# Contents

 Illustration Sources . . . . . . . . . . . . . . . . . . . . 7
 Preface . . . . . . . . . . . . . . . . . . . . . . . . . . . . . . 9
1. Development of Spikes . . . . . . . . . . . . . . . . . . 11
  Origin . . . . . . . . . . . . . . . . . . . . . . . . . . . . 11
  Continued Development . . . . . . . . . . . 15
  Engineering Spikes. . . . . . . . . . . . . . . . . 21
  Spike Length . . . . . . . . . . . . . . . . . . . . . 23
  Head Markings . . . . . . . . . . . . . . . . . . . 24
2. Spike Designs . . . . . . . . . . . . . . . . . . . . . . . . . 25
  Stevens Spike . . . . . . . . . . . . . . . . . . . . . 25
  Burden Spike . . . . . . . . . . . . . . . . . . . . . 27
  Golden Spike . . . . . . . . . . . . . . . . . . . . . 29
  Calvert Spike . . . . . . . . . . . . . . . . . . . . . 33
  Corydon Winch Spike . . . . . . . . . . . . . 35
  Common Cut Spike . . . . . . . . . . . . . . . 35
  Goldie Spike . . . . . . . . . . . . . . . . . . . . . 41
  Long Shank Spike . . . . . . . . . . . . . . . . 43
  Cattle Guard Spike. . . . . . . . . . . . . . . . 45
  Diamond Spike . . . . . . . . . . . . . . . . . . . 45
  Dog Spike . . . . . . . . . . . . . . . . . . . . . . . 46
  Greer Spike . . . . . . . . . . . . . . . . . . . . . . 49
  Daly Spike . . . . . . . . . . . . . . . . . . . . . . . 50
  Bayonet Spike . . . . . . . . . . . . . . . . . . . . 50
  Faries Spike . . . . . . . . . . . . . . . . . . . . . . 51
  AREA Common Spikes . . . . . . . . . . . . 52
  Screw Spikes . . . . . . . . . . . . . . . . . . . . . 55
  Sessler Spike . . . . . . . . . . . . . . . . . . . . . 59
  Compression Spike . . . . . . . . . . . . . . . . 59
  Gage Lock Spike . . . . . . . . . . . . . . . . . . 60
 Bibliography . . . . . . . . . . . . . . . . . . . . . . . . . 61

# Illustration Sources

FIGURE
1-3 Smithsonian Institution photographs* of spikes in their collection. Photograph Nos. 51,494; 737373-9; 51,493.
4 David Stevenson, *Sketch of Civil Engineering of North America* (London, 1838).
5 Smithsonian Institution collection (author's photograph).
6 Courtesy of United States Steel Corporation. Copyright 1964.
7 Smithsonian Institution collection (author's photograph).
8 Smithsonian photograph No. 737373-13.
9-10 Photograph of spike and letter in author's collection.
11 Drawn by author.
12-13 Smithsonian photograph Nos. 737373-8; 737373-4.
14 Smithsonian Institution collection (author's photograph).
15 *American Railroad Journal* (New York, July 23, 1859).
16 Courtesy of Stanford News and Publications Services, Stanford University, Stanford, California. Photograph No. 6572-5.
17-18 Courtesy of National Park Service, Golden Spike National Historic Site, Utah. Collection Nos. 114 and 101.
19 Courtesy of Stanford University Museum of Art, Stanford, California. Photograph No. 20007.
20 Smithsonian photograph No. 78-4053.
21 John Ashcroft, *Railway Directory* (New York, 1868).
22 Poor's *Manual of Railroads of the United States* (New York, 1883).
23 Baltimore and Ohio Railroad, *Standard Plans for Maintenance of Way and Construction* (Baltimore, 1907). Courtesy of the Akron Railroad Club, Inc., Akron, Ohio.
†24 *Railroad Gazette* (New York, December 4, 1885).
†25 *Pocket List of Railroad Officials* (New York, 4th Quarter 1896).
†26 Walter Webb, *Railroad Construction* (New York, 1900).
†27 *Pocket List of Railroad Officials* (New York, 4th Quarter 1904).
†28-33 William C. Willard, *Maintenance of Way and Structures* (New York, 1915).
†34-35 Gregg Company, *General Catalogue 550* (Hackensack, N.J., 1920).
†36 C. Weiss, *Practical Railway Maintenance* (New York, 1923).
†37-38 Author's collection, drawn by author.

*All Smithsonian photographs courtesy of the Smithsonian Institution.
†Redrawn by author from original illustrations.

†39 Composite of Fig. 123 from Walter Webb, *Railroad Construction*, which provided an overall view but included no dimensions, and Fig. 22 from W. M. Camp, *Notes on Track Construction* (Chicago, 1903), which provided dimensions for the point only.
†40 E. E. Russell Tratman, *Railway Track and Track Work* (New York, 1901).
†41 New York Central and Hudson River Railroad, *Standard Plans Book*, (1907).
†42-45 Willard, *Maintenance of Way and Structures*.
†46 Tratman, *Railway Track and Track Work* (1901).
47 Author's collection, drawn by author.
48 *Pocket List of Railroad Officials* (New York, 4th Quarter 1904).
49 Author's collection, drawn by author.
†50 AREA, *Proceedings of the Eighteenth Convention* (1917).
†51 Simmons-Boardman, *Maintenance of Way Cyclopedia* (Chicago, 1921).
52 Author's collection, drawn by author.
†53-54 Smithsonian Institution collection, drawn by author.
55 Poor's *Manual of Railroads of the United States* (New York, 1896).
†56 Winter Wilson, *Elements of Railroad Track and Construction* (New York, 1915).
57 *Railway Review* (Chicago, November 7, 1885).
58 Author's collection, drawn by author.
†59 AREA, *Proceedings of the Nineteenth Convention* (1918).
†60 AREA, *Proceedings of the Twenty-Second Convention* (1921).
†61 AREA, *Proceedings of the Thirty-Eighth Convention* (1937).
†62 AREA, *Proceedings of the Fortieth Convention* (1939).
†63 Bethlehem Steel Corporation, *Railroad Fasteners Booklet No. 3193* (Bethlehem, Pa., June 1977).
†64 Smithsonian photograph No. 737373-12.
65 Ashcroft, *Railway Directory*.
†66 Tratman, *Railway Track and Track Work* (1901).
†67 W. Kendrick Hatt, "Holding Force of Railroad Spikes in Wooden Ties," *U.S. Department of Agriculture Forest Service Circular 46* (Washington, D.C., December 26, 1906).
†68-69 Willard, *Maintenance of Way and Structures*.
†70 William H. Sellew, *Railway Maintenance Engineering* (New York, 1919).
†71 Bethlehem Steel, *Railroad Fasteners Booklet No. 3193*.
†72 *Railway Maintenance Engineer* (Chicago, December 1919).
†73 Simmons-Boardman, *Railway Engineering and Maintenance Cyclopedia* (Chicago, 1939). Dimensions are from spike in author's collection.
†74 Bethlehem Steel, *Railroad Fasteners Booklet No. 3193*.

†Redrawn by author from original illustrations.

# Preface

The excitement of discovering railroading history has been the most rewarding part of researching and collecting railroad spikes. The early days of railroading in America saw many interesting experiments in track and equipment construction—some successful, some not. Railroad spikes go back to these very early days—first being used successfully in the 1830s. Thereafter, the continuing development of railroads through the eighteen and nineteen hundreds brought many innovations and improvements. Railroad spikes share in these changes providing a rich variety of designs and shapes. Researching spikes naturally leads to discovery of local railroad history as well as history of entire railroad systems—all part of the excitement of spike collecting.

This book is not intended to be a complete history of railroad spikes nor a detailed engineering essay. Rather, it is only intended that the book be a stimulant to the collecting and researching of information about spikes.

Books have provided some background on spikes. Those written in the 1800s gave sketchy discussions on spike sizes and use. The flurry of railroad books written between 1900 and 1930 gave more detailed engineering accounts of the spikes in use. Since then, few railroad engineering books have been written. Nonetheless, books have provided a starting point for researching the history of spikes.

The source of most of the information about spikes is going to have to be obtained from interested collectors researching local railroads and steel manufacturers. There has been too little written and too many manufacturers and users of railroad spikes during the last 150 years to be researched by one individual.

Time is against us because, since 1937, the design of the railroad spike has been standardized to the American Railroad Engineering Association design, and, as far as I know, most spike manufacturers are making spikes which conform to this design. As track is respiked or ripped up, the old spikes are readily sold as scrap iron and, thus, forever lost. Remaining spikes are rusting away and the chance of discovering a premium quality spike is becoming more remote.

Abandoned rail lines are an excellent source of spikes. If the track has been ripped up, a metal detector can help find buried spikes. Periodically, railroads rebuild their existing track. Old spikes may be found discarded to the side of the track or still driven in old ties.

Antique stores and flea markets offer a potential source, although spikes have not yet become a big-selling item. Junk dealers may also have some old spikes.

Most railroad museums have spikes in their collections, saved from the historic past. The Smithsonian Institution also has many very old spikes. But, perhaps the richest untapped source of spikes is private collections. People tend to save odd-looking or old objects.

Collecting is not limited to spikes from standard gauge railroads. Small amusement park railroads, mine trackage, street railways, narrow gauge railroads, subway lines, elevated lines, and miniature model railroads have all used spikes, expanding the number, sizes, and shapes available. To sum it all up, if it holds rail down, it's collectible.

If you are at all interested in railroading, collecting spikes is the perfect hobby. Spikes are relatively inexpensive. There were plenty of them — about 10,000 per mile of track; and they reflect the history of American railroading, adding to the interest of collecting. Spikes have also been used on most foreign railroads at one time or another, opening up a whole new world of international collecting. For me, it has become a fascinating hobby.

\*   \*   \*

I am particularly indebted to my wife Jean, Mr. and Mrs. J. P. Joyce, Jr., of Oak Park, Illinois, and Mr. and Mrs. A. E. Farnham of Cape Cod, Massachusetts, who encouraged my interest in railroading and supported my preparation of this book.

I would also like to acknowledge the Smithsonian Institution; Lancaster County Library; Bethlehem Steel Corporation; United States Steel Corporation; American Railway Engineering Association; Stanford University Museum of Art; Union Pacific Railroad Company; National Park Service, Golden Spike National Historic Site, Utah; all those who contributed spikes to my collection; and everyone who helped me in the research and preparation of *Railroad Spikes: A Collector's Guide*.

# 1

# Development of Spikes

## *Origin*

America was a young, thriving country when the first railroads were built in the 1820s and 1830s. Important towns and markets were being linked by these new railways providing the beginning of a fast, efficient transportation system for commerce.

In 1830, Colonel John Stevens, a prominent railway pioneer and early advocate of railways, headed a movement which secured a charter from the New Jersey legislature for the Camden and Amboy Railroad. The railroad would eventually connect Amboy, New York, and Camden, New Jersey. In September, 1830, the board of directors of the new railroad ordered Robert L. Stevens, the president and chief engineer and son of Colonel John Stevens, to go to Britain to inspect and report on railroad construction there and to buy rail. The British had pioneered commercial railways in the 1820s, and it was common practice for engineers to study the British lines.

At the time, strap rail, one of the earliest forms of rail, was still in use in America and Europe. This rail was simply flat pieces of iron strap (fig. 1) laid on longitudinal stone or wooden stringers and held in place by small nails with countersunk heads (fig. 2). Strap rail eventually proved unsatisfactory because the iron strap would occasionally break loose, slash through the bottom of cars, and injure riders.

Another rail in wide use in both Europe and America was the British edge rail (fig. 3), formed with a thin web[1] and enlargement at one end. The rail was inserted into and supported by iron chairs nailed to wooden plugs in stone blocks. Both iron and stone blocks were very expensive at the time.

---

1. The parts of a rail are illustrated in figure 11.

FIGURE 1. Above, *Strap rail used on the Baltimore and Ohio Railroad, 1829.*
FIGURE 2. Below, *Nail used in strap rail, Camden and Amboy Railroad.* Dimensions: 3¾ inches by ⅜ inches. Smithsonian Collection No. 181,110.

The problems with these rail forms inspired Robert Stevens to design a unique shape while on the voyage to Europe. Francis B. Stevens, nephew of Colonel John Stevens, described the rail in a letter to James M. Swank, Special Agent to Statistics, who was compiling a census report of 1880 on iron and steel manufacturers. He wrote that the Stevens rail (fig. 4) was formed by ". . . adding the broad flange on the bottom, with a base sufficient to carry the load, and shaped so that it could be secured to the wood below it by spikes with hooked heads; thus dispensing with the cast-iron chair, and making

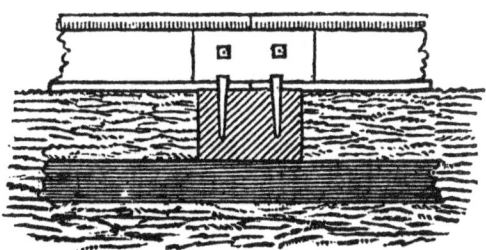

FIGURE 3. Above, *Edge rail on the Philadelphia and Columbia Railroad, ca. 1833.* FIGURE 4. *Construction of the Camden and Amboy Railroad in the 1830s;* left, *Stevens spike holding T rail;* below, *Fishplate joining rail.* The combination of Stevens T rail, hook-headed spike, and fishplate was the beginning of modern railroad construction.

the rail and its fastening such as is now in common use."[2] The new rail design weighing 36 pounds to the yard,[3] saved a large amount of iron by dispensing with the iron chairs, cutting costs considerably. Robert Stevens also invented the fishplate[4] for joining rail ends together. The method represented a great advance in the systems then in use.[5] The employment of ties, spikes, and fishplates, as illustrated in Stevenson's 1838 *Civil Engineering in America*, is shown in figure 4.[6]

Since the rail design was such a radical departure from contemporary manufacture, Robert Stevens had great difficulty in getting anyone in Britain to make the rail. Finally he persuaded Mr. Guest, a member of parliament and proprietor of a large iron works, to roll the rails at his iron works in Dowlais, Wales, but only upon promising to pay for any damages to the equipment. After several tries, the rail was successfully rolled. The first spikes were made in Britain also.[7] A spike from the Smithsonian Institution Collection identified as being used on the Camden and Amboy Railroad is shown in figure 5.

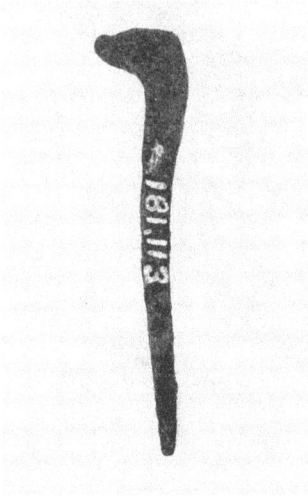

FIGURE 5. *Spike used on the Camden and Amboy Railroad. Dimensions: 6 inches by ½ inch square. Smithsonian Collection No. 181,113.*

The first shipment of rail arrived in Philadelphia in May, 1831, and track laying started shortly afterward. The Camden and Amboy Railroad officially opened October 9, 1832.

---

2. J. L. Ringwalt, *Development of the Transportation Systems of the United States* (Philadelphia, 1888), p. 86.
3. Originally the Stevens rail was called "H" rail, but by the 1840s it was commonly referred to as "T" rail. The first shipments of rail were 36 pounds, but later shipments were 42 pounds per yard.
4. Fishplates are also called joint bars, splice bars, angle bars, and angle plates.
5. Ringwalt, *Transportation Systems*, p. 86.
6. David Stevenson, *Sketch of the Civil Engineering of North America* (London, 1838), p. 247.
7. J. Elfreth Watkins, *The Camden and Amboy Railroad Origin and History* (Washington, D.C., 1891), p. 32. This is the only known reference to where the first spikes were made. More research is needed to conclusively prove this fact.

At first, the Stevens rail was spiked to stone blocks, but, during construction, there was a delay in getting stone blocks, so wooden crossties were substituted. It was quickly found that the track laid on wooden crossties rode better and held its gage better than track laid on stone blocks. Shortly thereafter, all the stone blocks were replaced with ties.

Without a doubt, the design of the Camden and Amboy Railroad using the Stevens rail, hook-headed spikes, and wooded crossties, was the first railroad in the world to be laid according to the present American practice.[8]

## *Continued Development*

The construction techniques used by the Camden and Amboy Railroad proved so successful that other railroads soon adopted the methods. H. S. Tanner in his book, *A Description of the Canals and Railroads of the United States,* published in 1840, gave the following account for the Boston and Lowell Railroad, "The usual form of spike, with a head projecting on one side only, is used to hold down the rail by overlapping its base on each side . . ." certainly a description of the Stevens spike.[9] According to Tanner, the Boston and Lowell Railroad, as well as the Providence and Stonington Railroad and the Long Island Railroad, underwent construction similar to the Boston and Providence Railroad which used 55 pound per yard rail and 9 ounce spikes that were ½ inch square and 6 inches long. Tanner also reported that the Philadelphia and Reading Railroad and the Washington branch of the Baltimore and Ohio used 45½ pound per yard Stevens rail with spikes 6 inches long with ¾ by ⅜ inch shanks and weighing ¾ pounds.

Other railroads converting to the Stevens rail by 1840 were the Philadelphia, Wilmington and Baltimore; portions of the Philadelphia and Columbia Railroad on which flat or bar-iron rails had originally been laid; on important New England roads including the second tracks of the Boston and Worcester and the Boston and Providence; New Castle and Frenchtown; and portions of the Georgia Railroad.[10]

---

8. J. Elfreth Watkins, *The Development of the American Rail and Track* (Washington, D.C., 1891), p. 670.
9. H. S. Tanner, *A Description of the Canals and Railroads of the United States* (New York, 1840), p. 37
10. Ringwalt, *Transportation Systems*, p. 86.

The Stevens T rail had to be imported until 1845 when the first T rail was rolled in America at the Montour Rolling Mill, Danville, Pennsylvania. The early American T rails were made of inferior iron. This was one of the causes that led to the adoption of the 1845 pear-shaped head (fig. 6) which provided more iron in the head area for added strength. The introduction of the Bessemer steel-making process in 1856 gave the structural strength needed to make stronger and larger rails with less body in the head. Thereafter, the size and shape of the rail steadily increased to the point that 155-pound rail is not uncommon.

Actually, the first rail made in America was iron cast in a U-form (fig. 6) and rolled at the Mount Savage Mill, Maryland, in 1844. It was used on the Baltimore and Ohio Railroad between Mount Savage and Cumberland, Maryland, but it never proved satisfac-

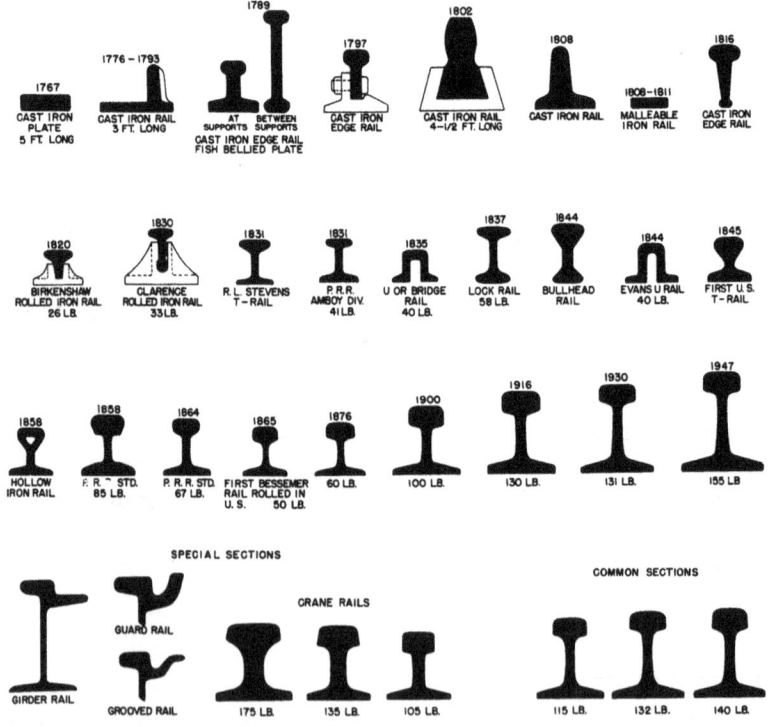

FIGURE 6. *Sketches of cross-sections of rail from the earliest periods of railroading until the present, showing the evolution of the modern railroad rail design.*

tory as a practical rail design. A spike used with the "U" rail is shown in figure 7.

Spikes, on the other hand, were being made in America since 1836 when Henry Burden, an inventive genius, developed a hook-headed spike-making machine.[11] Mr. Burden was superintendent of the Troy Iron and Nail Factory, which became the major supplier of spikes in these early days and for many years afterward.

FIGURE 7. *U rail spike, ca. 1855, found in the ground near Franconia, Virginia. Dimensions: 6 inches by 9/16 inches. Smithsonian Collection No. 322,483.*

From the 1830s to the 1920s, a time of experimentation and individuality in style existed. The simple hook-headed design took on a seemingly infinite variety of shapes and sizes. Unique designs such as the Greer, Goldie, and Dog enjoyed momentary popularity. Serious competition from the alternative screw spike really never developed. Some spikes were even handmade (fig. 8).

At the turn of the century, track construction was improved by adding tie plates between the rail and tie to prevent the rail from damaging the tie. As a result, spikes were made a little longer to compensate for the extra distance added by the tie plate.

Work started on standardizing the railroad spike design in 1911, and finally in 1918 spike manufacturers and railroad representatives agreed upon a standard spike design developed by the track committee of the American Railway Engineering Association (AREA). The AREA standard was further refined in 1921 and 1937 and, since that time, has remained the standard of the industry.[12]

---

11. Margaret Burden Proudfit, *Henry Burden His Life and History of His Inventions* (Troy, N.Y., 1904), p. 17.
12. AREA standardized rail design in 1915.

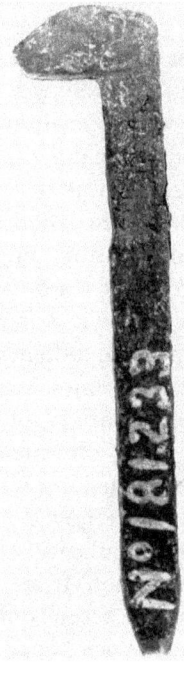

FIGURE 8. *Handmade spike from the South Carolina Railroad, made during the Civil War ca., 1860–1865. Dimensions: 6¼ inches by 9/16 inches. Smithsonian Collection No. 181,233.*

Throughout its long existence, the spike has retained the original "hook-headed" appearance of the 1830s, but this design has not always enjoyed the reputation as the best rail-tie fastening device. An article, "The Railroad Spike—An Anachronism of the Twentieth Century," appeared in the August 8, 1908, *Scientific American* and said in part:

> Although three quarters of a century has passed and each decade has seen a steady growth in the weight and speed of railroad trains and in the cost and quality of the tracks which carry them, we do not hesitate to say that in the whole field of Mechanical and Civil Engineering it would be impossible to find a device which is such an astonishing anachronism as the miserable little piece of ⅝ inch square iron known as the railroad spike of which it may be truly said that it has absolutely no other qualifications to recommend it beyond that it can be cheaply made and quickly driven into place.[13]

Well, three quarters of a century after these unkind words, the railroad spike has not been replaced as the common rail fastening

---

13. "The Railroad Spike—An Anachronism of the Twentieth Century," *Scientific American,* August 8, 1908, p. 86.

even though better fastenings have come along. The main reason was stated in the article. The spike is cheaply made and easily driven, thereby reducing the cost of constructing and maintaining the tracks.

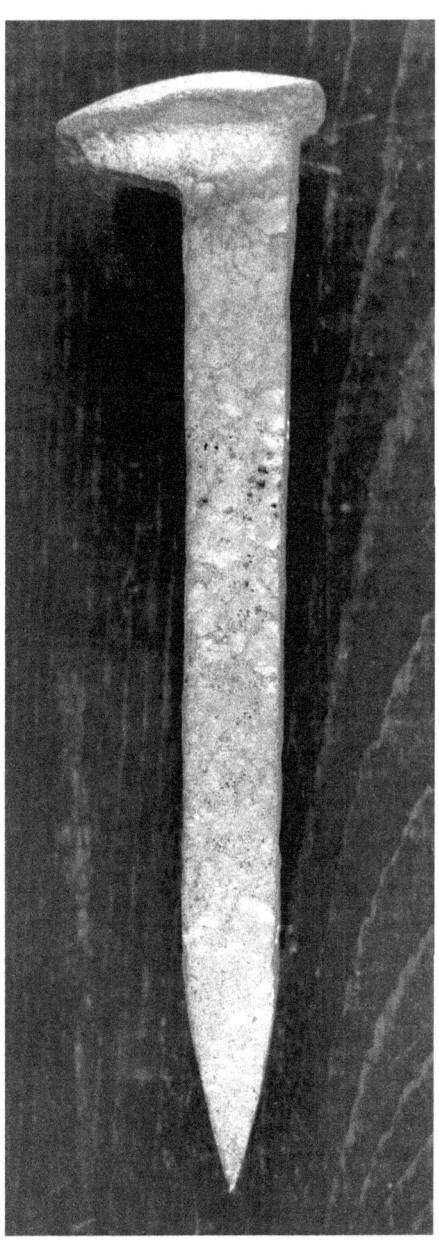

FIGURE 9. *Spike from the railroad connecting Launch Complex 5/6 and 26 at Cape Canaveral, Florida.*

Unfortunately, the demise of the spike still seems inevitable. Wood for ties is becoming scarce and expensive. Alternative means of constructing railroads are being developed mainly using concrete or steel ties. New designs such as the resilient and gage lock spikes are being tried with some success.

Looking back, the era of the railroad spike has enjoyed some golden moments though. One proud moment occurred on May 5, 1961, when Alan Shepard lifted off from Cape Canaveral, Florida, in a Mercury capsule aboard a Redstone rocket to become the first American into space. His flight, a 15-minute, 115-mile high shot into space ended on target in the Atlantic Ocean. The gantry used to assemble that rocket rode on a railroad fastened with ordinary

**DEPARTMENT OF THE AIR FORCE**
HEADQUARTERS AIR FORCE EASTERN TEST RANGE (AFSC)
PATRICK AIR FORCE BASE, FLORIDA 32925

OFFICE OF
THE COMMANDER

HISTORY OF RAILROAD SPIKE

The spike on this plaque was removed from the railway connecting Launch Complex 5/6 and 26, now the site of the Air Force Space Museum, Cape Kennedy Air Force Station.

This railway was used to move the gantry away from the launch pad prior to lift off. At the time these launch complexes were active, the missiles were assembled on the launch pad by use of the service tower.

Complex 5/6 and the railway are of historical significance in the United States Space Program. This complex was the site of the launch of EXPLORER I, the first U.S. satellite, and Alan B. Shepard's historic fifteen minute, sub-orbital flight on 5 May 1961.

DAVID M. JONES
Major General, USAF
Commander

FIGURE 10. *Letter explaining the significant role the railroad spike played in Alan B. Shepard's historic fifteen-minute, suborbital flight on May 5, 1961. Since the letter was written, Cape Kennedy Air Force Station, which is adjacent to NASA Kennedy Space Center, has been renamed Cape Canaveral, the original name.*

AREA railroad spikes (fig. 9). The Commander of the Air Force Eastern Test Range, which had responsibility for the launch complex facilities, prepared a letter (fig. 10) as testimony to the importance that the railroad spike played in this momentous occasion.

The most memorable moment, though, has to be the driving of the "Golden Spike" into a polished California laurel tie on May 10, 1869, symbolizing the joining of America, east with west, as the Central Pacific and the Union Pacific railroads came together at Promontory, Utah. One hundred years later as a part of the National Park System, the Golden Spike National Historic Site was dedicated there, thus enshrining for all time the significant role that the railroad spike has played in the development of America.

## Engineering Spikes

Spikes look the way they do because they must:
 a. Hold the rail down
 b. Keep the track to gage[14]

There are many forces which test the ability of the spikes to perform these functions. Track is subjected to severe vertical forces. The rail is first forced down as the train wheels pass, then, springs back up from the natural reaction of the rail. This continual wave motion gradually works to pull the spike up and out of the hole. At the same time, the rails are being forced outwards by the wheels, due to lateral pressure and blows of the wheel flanges against the rail heads. This lateral force works the spike back and forth in the hole, loosening the spike, damaging the tie, widening the gage, and, perhaps, necking (wearing away the neck) the spike.[15]

On curves there is a tendency for the rail to overturn. The rail tilts on its outer edge, draws up the inner spike or twists its head and forces the outer spike into the wood. Tie plates help resist the lateral force by making the spikes work together to aid in reducing the overturning tendency.

Actually, spikes are well engineered to resist these destructive forces by using a simple design (fig. 11) divided into three areas: head, shank, and point. Each area has a specific role to play in holding the rail down and keeping the track to gage which is worthy of further explanation.

---

14. Charles Weiss, *Practical Railway Maintenance* (New York, 1923), p. 118.
15. E. E. Russell Tratman, *Railway Track and Maintenance* (New York, 1926), pp. 92-93.

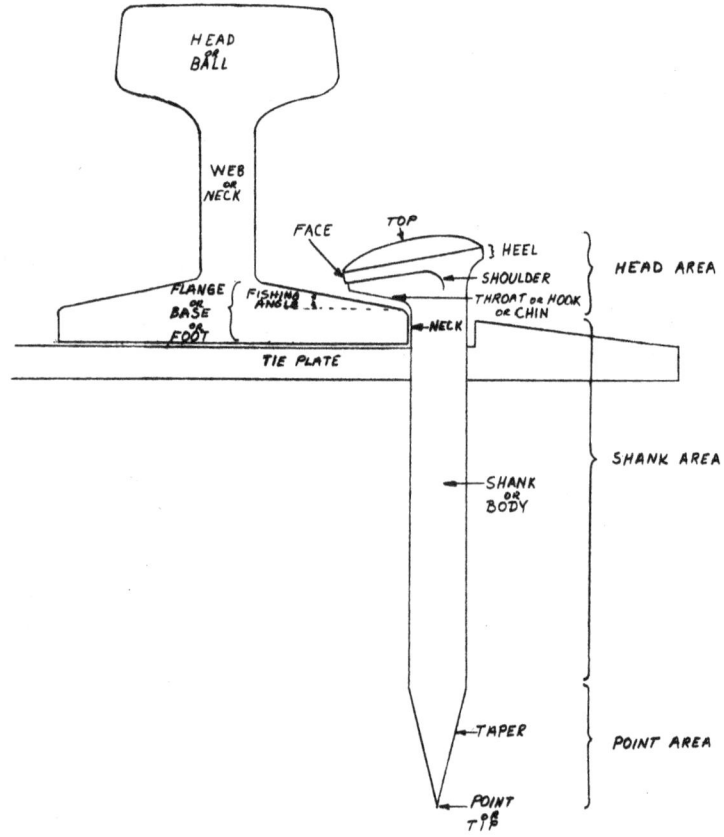

FIGURE 11. *Spike and rail parts nomenclature.*

1. Spike Head Area
   The railroad spike is often called a "hook-headed" spike because of the hooked shape of the head which bears against the rail base. Besides holding the rail down, the spike head provides a means of driving the spike into the tie, gripping and pulling the spike out, and holding tie plates down.
2. Spike Shank Area
   The holding power of a spike comes from the resistance between the depressed wood fibers and the shank which extends from just under the head to the beginning of the point. When first driven, the wood fibers are depressed and pushed down. As the spike is subjected to vertical forces tending to

lift the spike out of the tie, the wood fibers return to their original position increasing the hold on the spike.
3. Spike Point Area
As a spike is driven, the point cuts the wood fiber and diagonally depresses the fibers which bear against the shank and hold the spike in place. Attempts have been made to design points that improve the holding power of spikes by adding barbs, indentations, or special tapers. Most exotic point designs, however, have proved either difficult or expensive to manufacture and, therefore, never came into common use. The simple chisel point is the only point which has survived to become the standard of the railroad industry.

## *Spike Length*

The length of a spike is normally measured from under the head to the tip of the point. Some of the factors affecting the length are:
1. Type of tie wood (soft or hard). A half inch is added for soft wood.
2. Thickness of the tie. Spikes used on bridge timbers and stringers may be 1 to 2 inches longer.
3. Whether shims are used. Depending on the shim size, a 1-to 2-inch longer spike may be needed.
4. Thickness of the rail flange.
5. Size of the rail which is dictated by the speed and load of the trains.
6. Whether tie plates are used. One half inch is usually added for the plate.

In 1861, John Jervis, a prominent railroad civil engineer, noted that spikes are usually from ½ inch to ⅝ inch square and from 4½ to 5 inches long under the head.[16] W. B. Parsons reported, in 1886, that the most common spike was 5½ × 9/16 inches.[17] For narrow gauge (three-foot) railroads, W. Nicollis stated in 1878 that a 4½ × 7/16 inch spike does very well.[18] Today, a 6 or 6½ × ⅝ inch spike is used to hold the larger size rail needed for the heavier, faster trains being run.

---

16. John B. Jervis, *Railway Property* (New York, 1861), p. 130.
17. W. B. Parsons, *Track* (New York, 1886), p. 33.
18. W. J. Nicollis, *The Railway Builder* (Philadelphia, 1878), p. 108.

# Head Markings

It is common practice today to mark the head of the spike with a unique symbol identifying the manufacturer. The practice probably started in the early 1900s, about the time AREA added the requirement to its spike specifications.

Railroads seldom marked spikes with their railroad logo; however, the Delaware, Lackawanna and Western was a notable exception when it marked its screw spikes.

Some of the symbols that may be found on the head are listed below:[19]

| MARKING | DESCRIPTION |
|---|---|
| HC or C | Spike Made to High Carbon AREA Specifications for added strength. |
| CU | Spike made with copper added. Copper provides increased corrosion resistance.[20] |
| B‖S | Bethlehem Steel Corporation, Bethlehem, Pennsylvania. |
| CP & I | CP & I Steel Corp., Pueblo, Colorado. |
| DL & W | Delaware, Lackawanna and Western Railroad |
| ◇I | Inland Steel Corp., Chicago, Illinois. |
| J | Jones and Laughlin, Philadelphia, Pennsylvania. |
| RW | Rockwood Iron and Metal, Inc., Rockwood, Tennessee. |
| S | ARMCO Inc., Western Steel Division, Kansas City, Missouri. |
| T | United States Steel Corp., Pittsburgh, Pennsylvania. Originally used by the Tennessee Coal and Iron Co., which is now a division of U.S. Steel. |
| Y | Youngstown Steel, Youngstown, Ohio. |

---

19. This list is very incomplete. More research into the history of steel companies is definitely needed.
20. In areas handling a large number of "Brine Tank" refrigeration cars loaded with meat, fish, or like substances, track spikes are sometimes galvanized. This prevents deterioration from rust normally resulting from drippage from cars. A galvanized spike can be identified by its gray color.

# 2

# Spike Designs

## Stevens Spike

The Stevens spike, the first hook-headed spike, was invented in 1830 by Robert L. Stevens, President and Chief Engineer of the Camden and Amboy Railroad, for use with the Stevens T rail.

The only known illustration of the Stevens spike appeared in Stevenson's 1838 book, *Civil Engineering in America,* and was previously shown in Fig. 4. Unfortunately, no description was given, and the illustration is not a precise rendering.[1]

The Smithsonian has several spikes in its collection identified as being from the early days of the Camden and Amboy Railroad. These are described[2] in figures 12, 13, and 14.[3]

Only Smithsonian spike No. 180,796 (fig. 13) is positively identified as having been made by Guest and Co., Wales,[4] the same company that made the first Stevens T rail. No manufacturer is given for the remainder. In the 1830s, America certainly had the resources and capability of making the spikes either by hand or by machine. Continued research of these first spikes may yet uncover their origin.

---

1. Stevenson, *Civil Engineering,* p. 247.
2. Descriptions of each spike have been provided by the Smithsonian Institution.
3. The dimensions were calculated by the author and are approximate. In most cases, the spikes have suffered some deterioration from rust, making an exact estimation of original size impossible.
4. "J. E. Watkins, a former curator of the National Museum of History and Technology, Washington, D.C., attempted to write the history of the Pennsylvania Railroad in the early 1890s. Because it was an official project, he was given access to material then available—much of which has probably since been destroyed or lost by the railroad. From the information available to him, perhaps including original invoices, Mr. Watkins concluded that the first spikes were made by Guest and Co." Extract of letter from John H. White, Jr., Curator, Division of Transportation, Smithsonian Institution, to author on March 23, 1979.

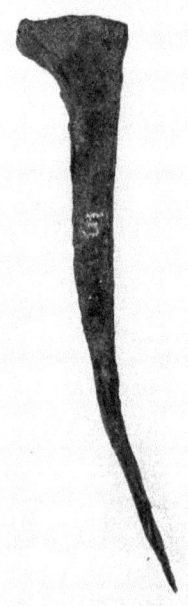

FIGURE 12. Above, left, *Spike used in the original construction of the Camden and Amboy Railroad at Hightstown, New Jersey, ca. 1830.* Dimensions: 6 inches long by ½ inch. Smithsonian Collection No. 180,824. FIGURE 13. Above, *First pattern with a hooked-head, made by Guest and Co., Wales, 1830.* Dimensions: 6 inches long by ⅜ inches. Smithsonian Collection No. 180,796. FIGURE 14. Left, *Early form of railroad spike, probably Camden and Amboy Railroad, 1833–1840. It was used to fasten the rails to stone blocks.* Dimensions: 6 inches long; shank, ⅝ by ⅜ inches. Smithsonian Collection No. 180,006.

## Burden Spike

The Stevens rail quickly came into general use in America by 1845 although several railroads continued to use other types of rail until they were worn out. One company that became a major supplier of spikes for the Stevens rail was the Troy Iron and Nail Factory of Troy, New York, originally formed in 1813. Henry Burden, the superintendent since 1822, was as inventive genius. One of his innovations was an 1834 patent for a machine that made countersunk railroad spikes for strap rail.[5] This patent was an improvement over his earlier May 26, 1825, patent for ship builder spikes.

In the winter of 1835–1836, Henry Burden visited England and, while there, learned that the flat strap rail would likely be superseded by the Stevens rail which needed a different spike. On his return home, he reconstructed his spike machine and began manufacturing the new hook-headed spikes within a short time. He made ten tons of the spikes for the Long Island Railroad as his first contract in 1836.[6] It was not until 1840 that he was awarded a new patent No. 1757 for the machine improvements. In the patent, Mr. Burden says about his new machine:

> The feeding of the rod, the cutting it off, and the pointing the spike are effected in the same way previously used by me . . . the heads of the hook or brad-headed spikes were, so far as I am informed, always made by hand; and they were necessarily imperfect, being deficient in that uniformity in shape and strength which are important requisites. My improvement of what I denominate a bending lever of some analogous device, by means of which the portion of the rod which is to constitute the head is bent down, so as to form an angle with the shank, and in then forcing up a heading die, properly formed, so as to upset[7] the bent portion and to cause it to assume the desired shape.

By 1848, Henry Burden had become sole owner and formed the Burden Iron Works. In 1864, the company was again reorganized as H. Burden and Sons.[8] By this time, the Burden spike was being pro-

---

5. Mr. Burden was awarded patent No. 8515X on December 2, 1834.
6. Proudfit, *Henry Burden*, p. 17.
7. UPSET means to mould the hot metal into the form of the spike head similar to the way a blacksmith works.
8. Samuel Rezneck, "Office Building 1881 Burden Iron Company, Troy," *A Report of the Mohawk-Hudson Area Survey* (Washington, D.C., 1973), p. 96.

FIGURE 15. *H. Burden & Sons advertisement of 1859.*

duced in three sizes: 6 × 9/16 inches, 5 × ½ inches, and 4 × 3/8 inches as shown in figure 15. Similar spikes were advertised by New York Railroad Chair Works[9] and the Corydon Winch Company[10] both of New York City.

Henry Burden died in 1871 leaving the firm to two of his four surviving sons. The two brothers were in constant disagreement which, together with a slowdown in Troy's role as an ironmaking center, led to the gradual decline of the Burden Company.

Incidentally, Henry Burden is probably more famous for his horseshoemaking machine patented in 1835; the unique machine quickly elevated Troy to the horseshoe capital of the world for many decades thereafter.

---

9. Henry V. Poor, *History of the Railroads and Canals of USA* (New York, 1860), p. 624.
10. James W. Low, *Railway Directory* (New York, 1861), p. xiv.

## Golden Spike

The joining of America, east with west, when the Union Pacific and Central Pacific came together at Promontory, Utah, was symbolized by the driving of the "Golden Spike" on May 10, 1869. This was a historic occasion for the nation and many dignitaries took part in the festivities.[11] There were four spikes presented: two gold, one silver, and one gold-silver-iron spike.[12]

FIGURE 16. *The "Golden Spike" which symbolized the joining of America in 1869, when the Union Pacific and Central Pacific railroads came together at Promontory, Utah.*

---

11. An exhaustive account of the occasion is given by J. N. Bowman in "Driving the Last Spike at Promontory 1869," reprinted in the *Utah Historical Quarterly*, Winter 1969, pp. 76–101.
12. Mr. Bowman provides a detailed history of each spike. His findings will not be repeated here.

## THE GOLDEN SPIKE

The "Golden Spike" (fig. 16) was made from a mould designed by William T. Garratt, a brass and bell founder of San Francisco.[13] Mr. Garrett most likely followed the pattern of a common spike from the Central Pacific Railroad (fig. 17).[14] Mr. Garratt also cast the

FIGURE 17. *Common spike used in the building of the Central Pacific Railroad ca. 1860. It was found on the old CPRR grade in the Golden Spike National Historic Site, Utah. Dimensions: 5 inches by ½ inch.*

---

13. Robin Lampson, "The Golden Spike is Missing," *The Pacific Historian*, Winter 1970, p. 20.
14. "We think it [Golden Spike] was precisely copied from common spikes used on Central Pacific Railroad. We are led to believe this because we have spikes in our collections which have been recovered from the old CPPR grade, and they are the mirror image of the last spike." Extract of letter from Paul L. Hedren, Interpretive Specialist National Park Service, Golden Spike National Historic Site, Utah., to author on December 28, 1981.

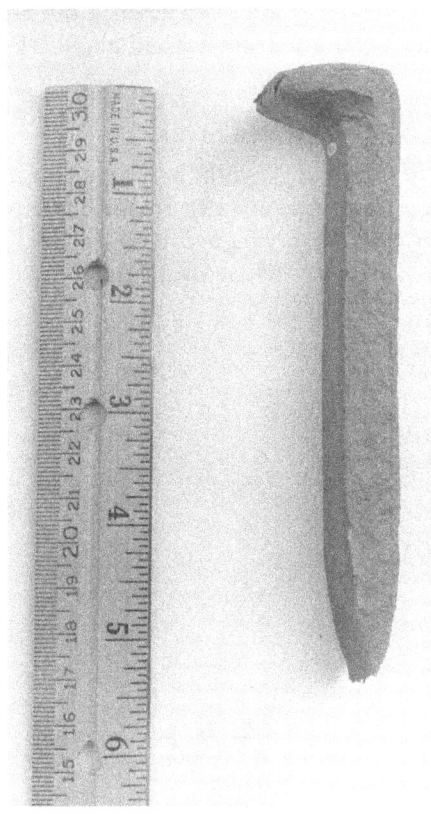

FIGURE 18. *Common spike from the Union Pacific Railroad, ca. 1860. It was found on the Union Pacific grade in the Golden Spike National Historic Site, Utah. Dimensions: 5 inches by ½ inch.*

spike, while the finishing and inscribing was done by Schulz, Fischer and Mohrig of San Francisco.[15] The spike presented by David Hewes, a wealthy contractor of San Francisco, was 5⅝ inches long (overall), 17/32 inches square, weighed 14.13 ounces, and contained 13.377 approximate gold or 17.6 carats finished. The Golden Spike is now on display at the Stanford University Museum of Art, Stanford, California. A common spike from the Union Pacific Railroad is shown in figure 18.

THE SECOND GOLD SPIKE

A second gold spike donated by Frank Marriott, owner of the *San Francisco News Letter,* was also driven and afterward given to General

---

15. Lampson, "The Golden Spike is Missing," p. 20.

G. M. Dodge, Chief Engineer of the Union Pacific. The spike was 5 inches in length and weighed about 9½ ounces. Its fate beyond this point is unknown.

### THE NEVADA SILVER SPIKE

The State of Nevada had a pure Virginia City silver spike made for the occasion (fig. 19). It was 6 inches long, ¾ inches square with a 1½ inch head, weighed 10½ ounces. This spike is now at the Stanford University Museum of Art.

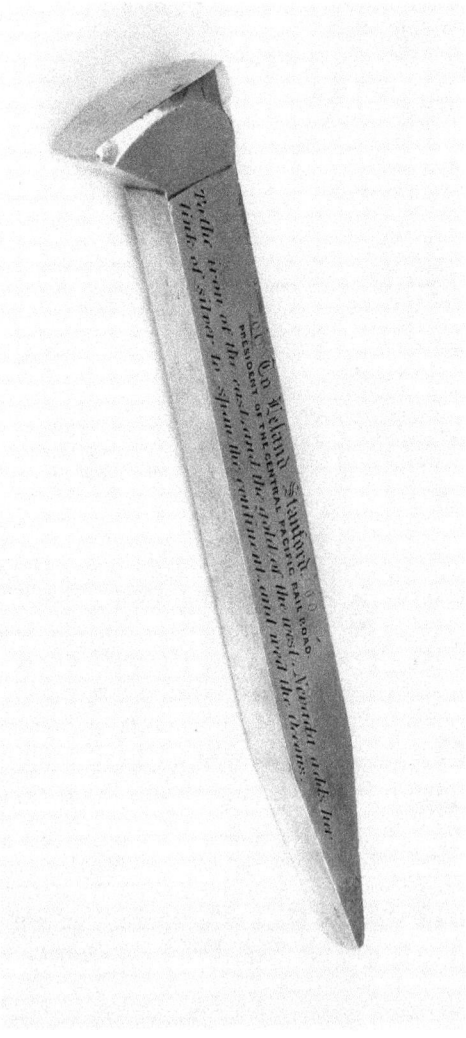

FIGURE 19. *Silver spike presented by the State of Nevada.*

## The Arizona Iron-Silver-Gold Spike

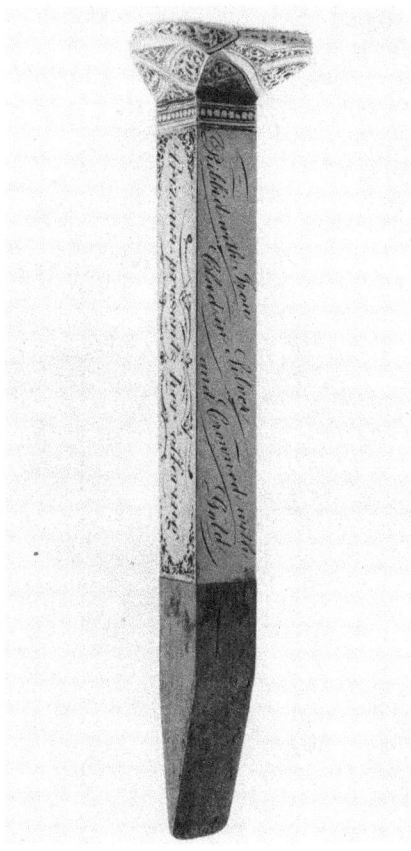

FIGURE 20. *The Arizona Iron-Silver-Gold Spike.*

A. K. P. Safford, then recently appointed territorial governor of Arizona, also presented a spike at the ceremony. The inscription on the spike sums up the significance of the event. It reads: "Ribbed with iron, clad in silver and crowned with gold Arizona presents her offering to the enterprise that has banded a continent, dictated a pathway to commerce. Presented by Governor Safford." The spike (fig. 20) is 6 inches in length, ¾ inches thick, weighs 10¼ ounces, and was prepared by D. W. Laird, a jeweler in San Francisco.

After the ceremony, Sidney Dillon, a member of the Board of Directors of the Union Pacific Railroad, acquired the spike. It remained in the family until an heir donated it to the Museum of the City of New York which indefinitely loaned it to the Smithsonian Institution where it is on display.

## Calvert Spike

The Calvert Iron and Nail Works of Baltimore featured their spike in an advertisement in the 1868 edition of Ashcroft's *Railway Directory* (fig. 21). Examples of spikes that closely resemble the Calvert spike have been uncovered from the historic Union Pacific Railroad grade in the Golden Spike National Historic Site.[16]

---

16. Letter from Paul L. Hedren, Interpretive Specialist, National Park Service, Golden Spike National Historic Site, to author on December 28, 1981.

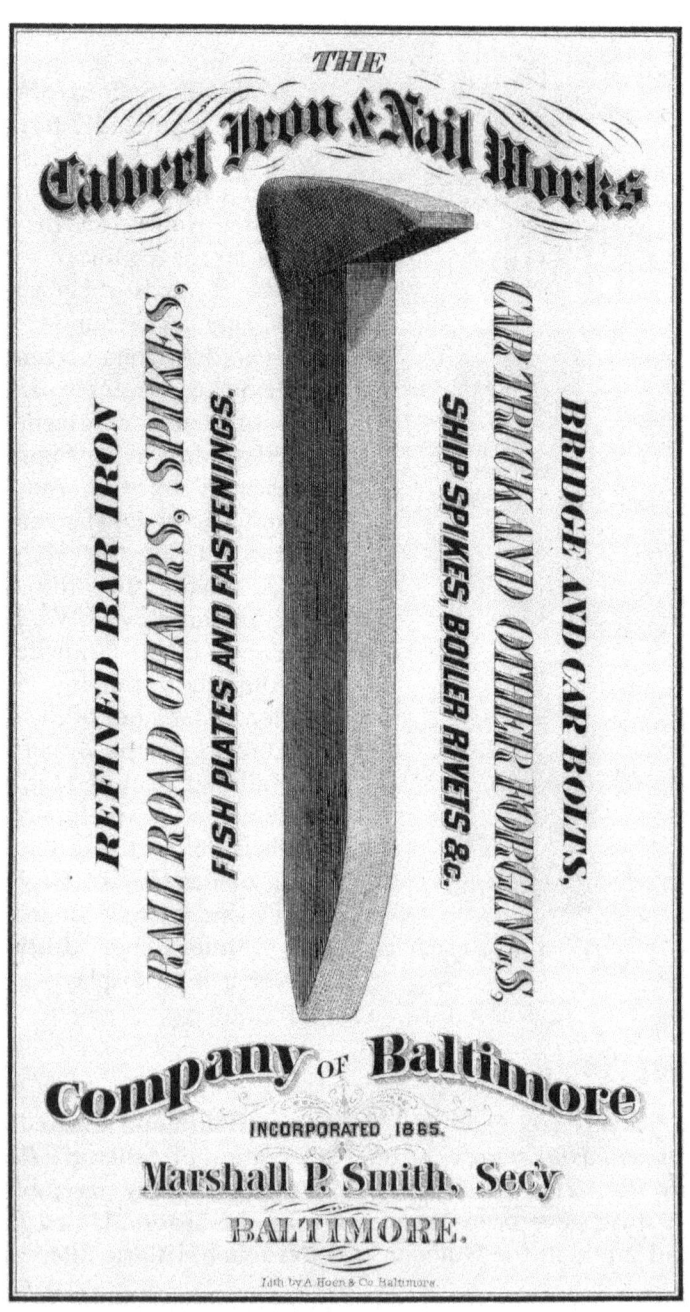

FIGURE 21. *Calvert Iron and Nail Works 1868 advertisement. The design was similar to the Burden shape.*

## Corydon Winch Spike

Corydon Winch Company was mentioned earlier as having sold spikes similar to the Burden design in the 1860s. By the 1880s, the company, now of Philadelphia, changed the design to that illustrated in figure 22. The new design is unique because of its flat head.

FIGURE 22. *Corydon Winch advertisement of 1883. The flat head is unusual.*

## Common Cut Spike

The hooked-head, square shank, and chisel point characterizes the common cut spike. The Baltimore and Ohio spike of 1906 (fig. 23) typifies this spike. These simple spikes were made by the millions during the late 1800s and early 1900s.[17] They have no distinguishing marks and come in a seemingly infinite number of sizes and head shapes. They are the spikes that everyone thinks of as the "ordinary" railroad spike. By 1918, the design was refined and standardized into the American Railway Engineering Association (AREA) spike which will be discussed later.

Examples of some common spikes are shown in figures 24–38.

---

17. More than fifty manufacturers were making spikes at the turn of the century as compared to a dozen today.

35

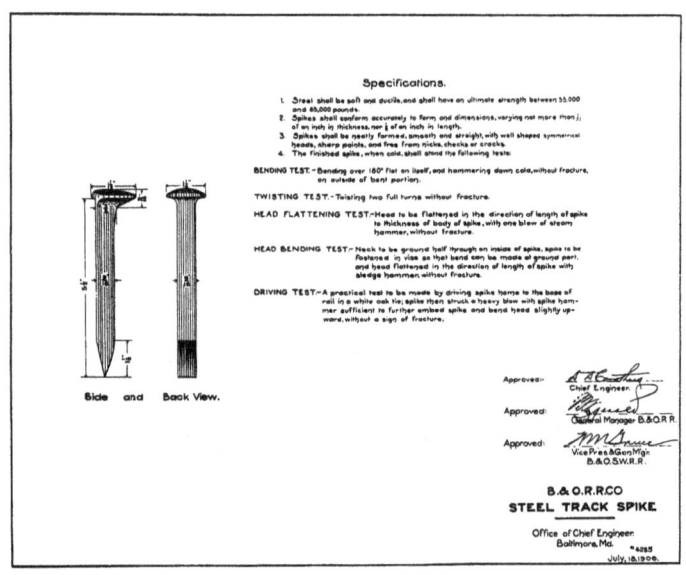

FIGURE 23. Above, *Specifications and design of standard track spike for the Baltimore and Ohio Railroad, 1906.* FIGURE 24. Below, left, *Standard spike, Pennsylvania Railroad, ca. 1885.* FIGURE 25. Below, right, *Common cut spike, ca. 1896, made by Central Iron and Steel Co., Brazil, Indiana.*

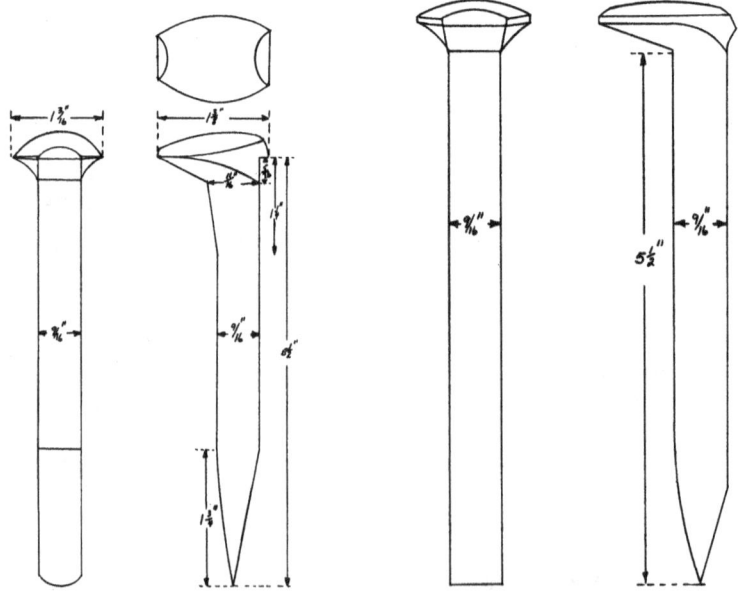

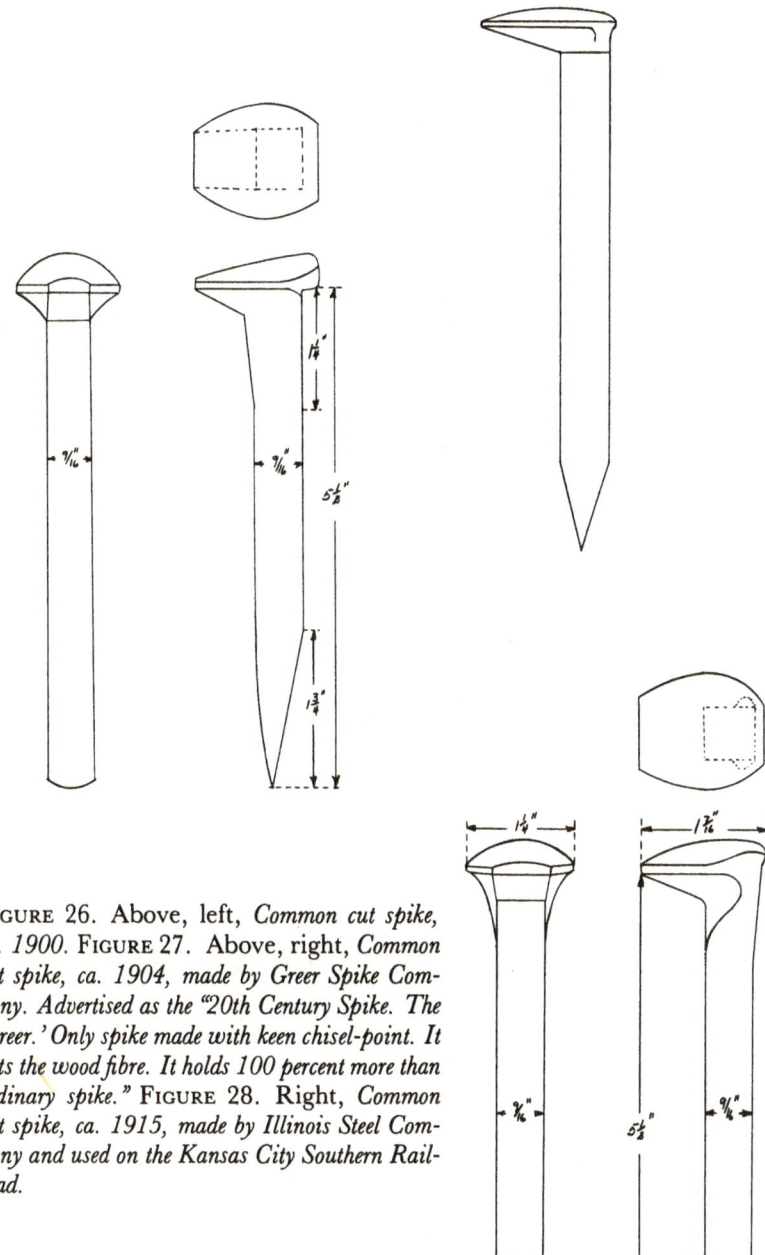

FIGURE 26. Above, left, *Common cut spike, ca. 1900.* FIGURE 27. Above, right, *Common cut spike, ca. 1904,* made by Greer Spike Company. Advertised as the "20th Century Spike. The 'Greer.' Only spike made with keen chisel-point. It cuts the wood fibre. It holds 100 percent more than ordinary spike." FIGURE 28. Right, *Common cut spike, ca. 1915,* made by Illinois Steel Company and used on the Kansas City Southern Railroad.

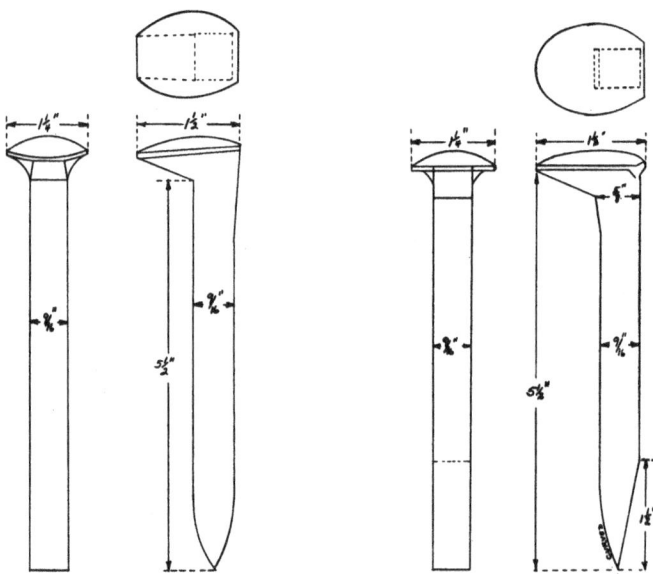

FIGURES 29–32. *Common cut spikes, ca. 1915, used on the Canadian Government Railways* (above left), *Baltimore and Ohio Railroad* (above right), *Pennsylvania Railroad* (below left), *and New York Central and Hudson River Railroad* (below right), *respectively.*

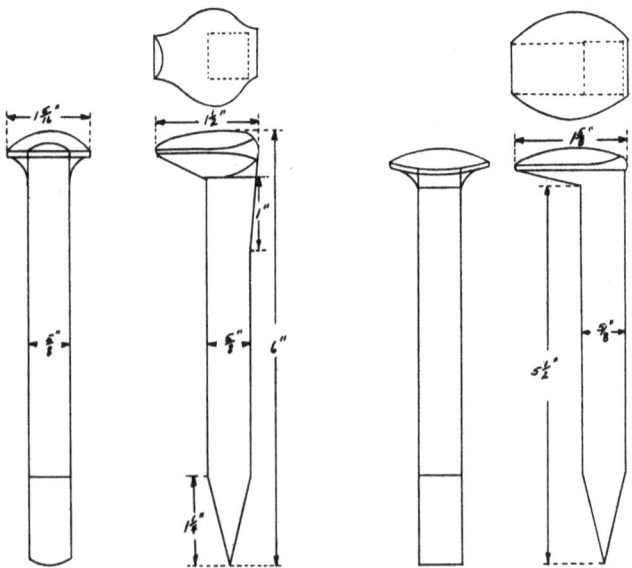

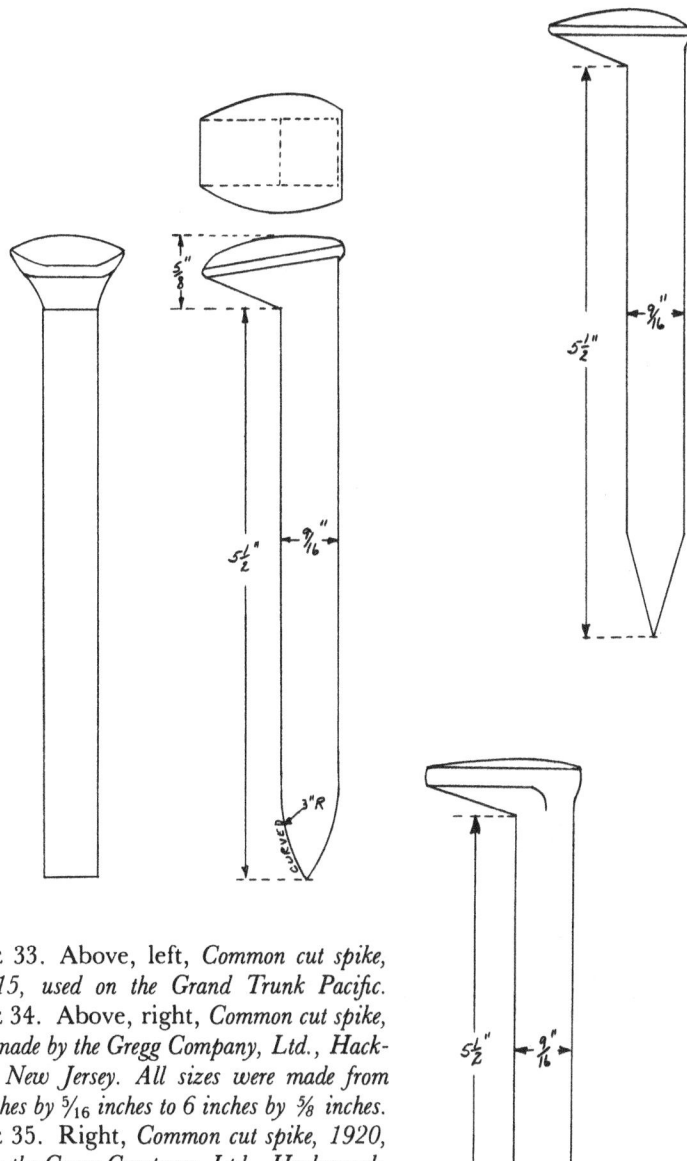

FIGURE 33. Above, left, *Common cut spike, ca. 1915, used on the Grand Trunk Pacific.*
FIGURE 34. Above, right, *Common cut spike, 1920, made by the Gregg Company, Ltd., Hackensack, New Jersey. All sizes were made from 2½ inches by 5/16 inches to 6 inches by 5/8 inches.*
FIGURE 35. Right, *Common cut spike, 1920, made by the Gregg Company, Ltd., Hackensack, New Jersey. All sizes were made from 2½ inches by 5/16 inches to 6 inches by 5/8 inches.*

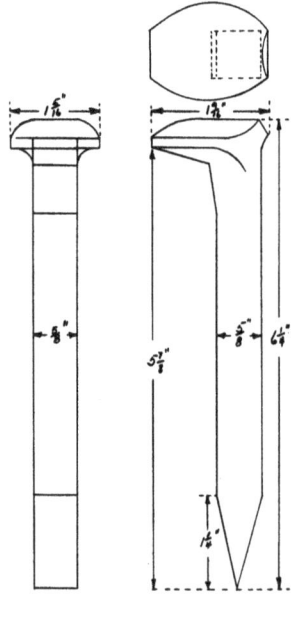

FIGURE 36. Left, *Common cut spike, ca. 1923,* used on the Union Pacific Railroad. FIGURE 37. Below, left, *Common cut spike, ca. 1890,* used on the Oxford and Southern Railroad, a narrow gauge railroad that ran between Oxford and Quarryville, Pennsylvania. FIGURE 38. Below, right, *Common cut spike, ca. 1910,* used on a small amusement park railroad near Gettysburg, Pennsylvania.

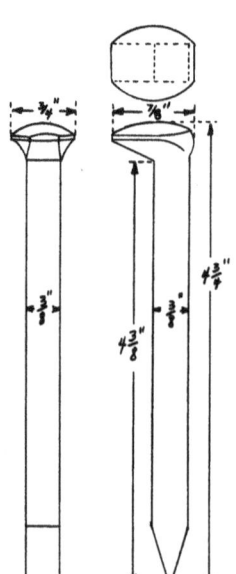
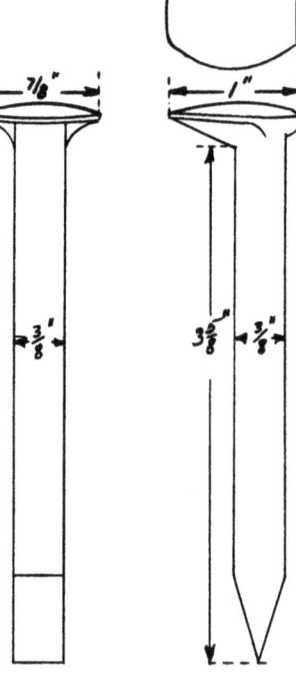

## Goldie Spike

William Goldie of West Bay, Michigan, was the holder of a number of patents, the most important being No. 413,342 titled "Method of Pointing Spikes" dated October 22, 1889. The patent illustrates several point designs that his machine could make. One of the designs is the common chisel point (figure 5 in the patent), but the most important is the Goldie point (figure 7 in the patent). In his 1908 book *Railway Track and Track Work*, Tratman says: "In the

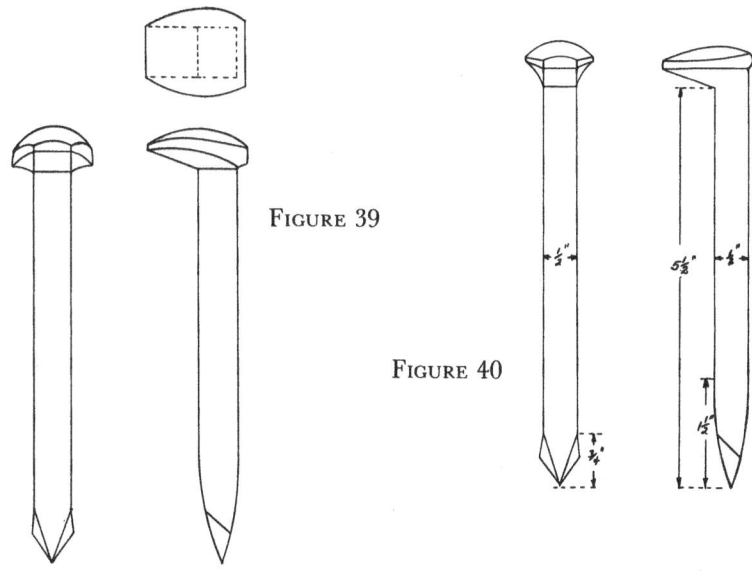

FIGURE 39

FIGURE 40

*Goldie spikes, ca. 1900 and 1901, respectively.*

Goldie spike the end is ground to a sharp point instead of to a chisel edge, while both edges and faces of the point are inclined so as to increase the cutting and wedging effect."[18] Superiority of the Goldie spike (figures 39 and 40) was based on the claim that the point was supposed to minimize the destruction of the wood fibers, thereby increasing the holding power.[19]

In his revised book of 1926, Tratman says: "The Goldie spike, with sharp diamond point, was used extensively at one time but has

---

18. E. E. Russell Tratman, *Railway Track and Track Work* (New York, 1908), p. 90.
19. Marshall M. Kirkman, *Building and Repairing Railways* (New York, 1903), p. 215.

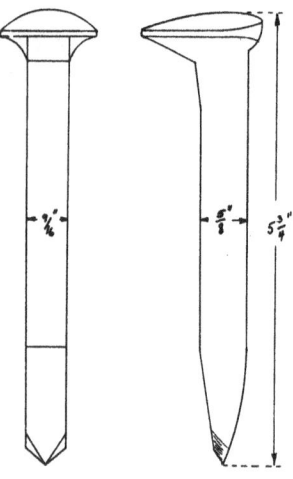

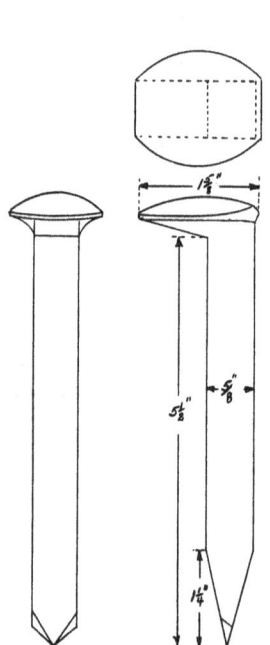

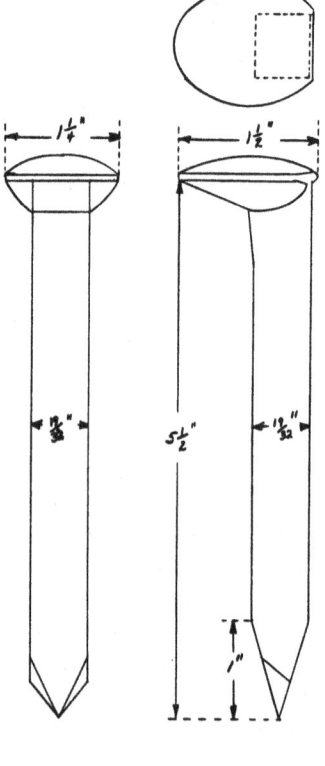

FIGURE 41. Left, *Goldie-type spike, ca. 1907, used on the New York Central and Hudson River Railroad.* FIGURE 42. Below, left, *Goldie-type spike, ca. 1915, used on the New York Central and Hudson River Railroad.* FIGURE 43. Below, right, *Goldie spike, ca. 1915, used on the Boston and Maine Railroad.*

declined on account of the extra cost."[20] The era of the Goldie spike ran roughly from the early 1890s through the 1920s, a period of thirty years.

Around 1900, a spike with a point that closely resembled the Goldie point was used by the New York Central and Hudson River Railroad as the standard for soft Carolina pine ties.[21] The spike was patterned after the standard spike of the Pennsylvania Railroad except for the Goldie-style tip.[22] Figure 41 shows a design from the 1907 New York Central Standard Plans Book. A later design in use in 1915 is shown in figure 42.

The Goldie spike enjoyed some popularity. In 1916, AREA compiled a list of railroads using the Goldie spike. The railroads were: Maine Central; Cleveland, Cincinnati, Chicago & St. Louis; Boston and Albany; Virginian; and the Boston and Maine (fig. 43).[23]

## Long Shank Spike

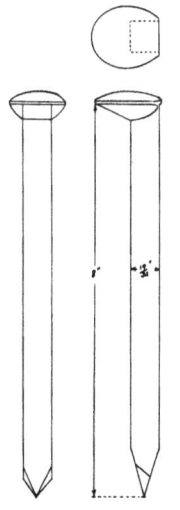

FIGURE 44. *Shim spike, ca. 1915, used on the Boston and Maine Railroad.*

In winter when the ground is frozen, the ballast, unless perfectly porous, is unequally heaved by frost. This causes unevenness in the rail surface since the rail is high in some places and low in others. On well-maintained track, a flat wood shim of the proper thickness to make the two rails level is inserted between the tie and rail on the low spots. Because of the added distance the shim places between the rail and the tie, shimming spikes (fig. 44) are used. These spikes are ordinary spikes, but their shanks are an inch or two longer. Some spikes

---

20. Tratman, *Railway Track and Maintenance* (1926), p. 96.
21. Winter Wilson, *Elements of Railroad Track and Construction* (New York, 1915), p. 66.
22. W. M. Camp, *Notes on Track Construction and Maintenance* (Chicago, 1903), p. 126.
23. American Railway Engineering Association, *Proceedings of the Seventeenth Convention* (Chicago, 1916), pp. 378-380.

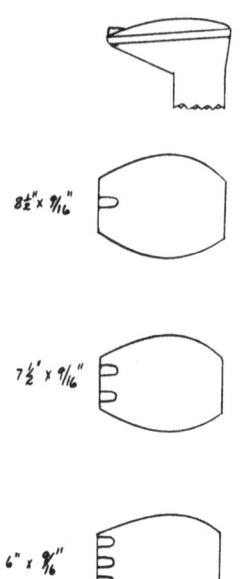

have small projections (fig. 45) which are used to quickly tell the section worker the length of the spike. When the ground thaws, the shims are removed along with the shim spikes and replaced with ordinary spikes.

Longer shanked spikes are occasionally used on bridges where the crossties are bridge beams which are considerably thicker, offering the opportunity for additional holding power that a longer spike provides.

FIGURE 45. Above, *Shim spike head markings, ca. 1915, used on the Canadian Government Railways. Markings indicate size of spike to section workers.*

At switches, longer spikes are sometimes used to hold down the switch stand on the headblock tie.[24] In 1885, the standard headblock spike used on the Pennsylvania Railroad was similar to the spike in figure 24 except it was 7 inches long under the head.[25] At the turn of the century, the standard headblock spike for the Pennsylvania Railroad was similar to its standard spike except it was 7 inches long (fig. 46).[26]

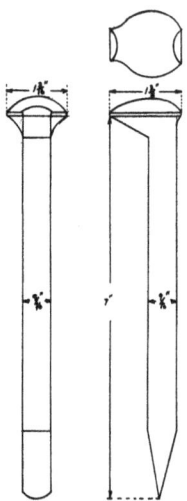

FIGURE 46. Above, *A headblock spike used on the Pennsylvania Railroad, ca. 1901. Note the rolled point which minimized the edge fins which can cause a spike to turn as it is driven.*

---

24. Headblock ties are longer ties on which the switch stand is fastened.
25. "Pennsylvania Railroad Standard Spike," *Railroad Gazette,* December 4, 1885, p. 771.
26. Camp, *Notes on Track Construction,* p. 126.

## Cattle Guard Spike

Occasionally, when workers lay track, the rail may have to be spiked to timbers which run in the same direction as the rails. This can happen with cattle guards or with bridge stringers. In such cases, special spikes may have been used which had the point turned 45 or 90 degrees from the standard (fig. 47). This design avoids timber splitting by allowing the the spike point to be driven across the wood grain.[27]

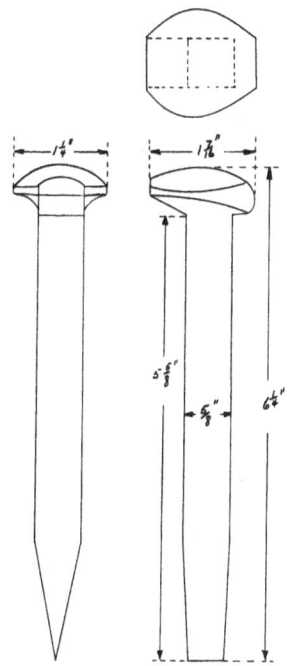

FIGURE 47. *Cattle guard spike. Note the point is turned 90 degress from normal.*

## Diamond Spike

A unique gouge point was used on spikes made around 1900 by the Diamond State Steel Company of Wilmington, Deleware. The Diamond spike, also called Diamond and Crescent spike,[28] has a gouge-shaped point. The face on the rail side of the point is convex, while on the back side of the spike the face of the point is grooved. The gouge shape was supposed to ease the entry of the spike and thereby minimize the destruction of wood fibers.[29] Sometimes, the point is ground to a sharp edge and may be convex for cedar ties and concave for oak or knotty wood ties.[30]

An advertisement appearing in *The Pocket List of Railroad Officials* in 1904 gives a good illustration of the spike (fig. 48). Another advertisement by the Diamond State Iron Co. in an 1894 publication

---

27. Ibid, p. 125.
28. E. E. Russell Tratman, *Railway Track and Track Work* (New York, 1901), p. 81.
29. "The 'Churchill' Rail Joint and 'Diamond' Spike," *Street Railway Journal*, Vol. XIV, No. 12, 1898, p. 812.
30. Tratman, *Railway Track and Track Work* (1901), p. 81.

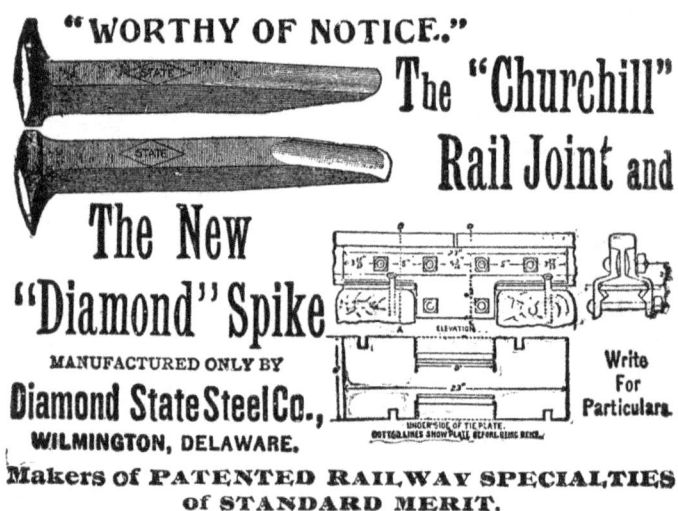

FIGURE 48. *Diamond spike 1904 advertisement. Note the gouge point.*

shows a similar shape to the gouge point except the gouge only extends about a quarter to a half an inch from the point. The tip of this spike is concave, forming two points which the advertisement claims will prevent the spike from turning when driven.[31]

## Dog Spike

The origin of the dog spike (the name coming from the resemblance to the snout and ears of a dog) lies in the rich history of European railroading. The design was popular in Europe at least back to the 1870s. In contrast, dogs[32] have been used only occasionally on American railroads. Reference to their use on American railroads first appears in books around 1900. Surprisingly, none of the many spike patents resemble the dog design. The dog spike (figs. 49-54) features unique side lugs which were intended for easy removal of the spike using a claw bar.

---

31. Road and Track Supply Association, *Complimentary Book*, (Chicago, 1894), p. 16.
32. The term "dogs" is a shortened name for a dog spike as used in John W. Barry's *Railways and Locomotives* (London, 1882), p. 104.

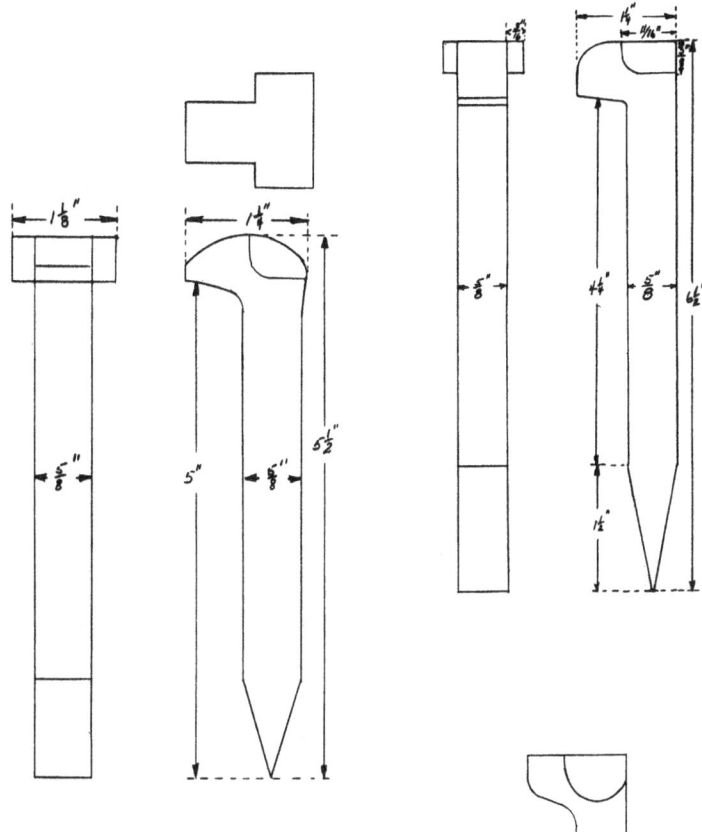

FIGURE 49. Above, left, *Dog spike, date unknown,* used on the Thailand Burma Railroad which is the famous World War II "Death Railroad" scene in the novel and movie Bridge over the River Kwai. FIGURE 50. Above, right, *Dog spike, ca. 1917,* an AREA design submitted at the 1917 convention as a matter of information. FIGURE 51. Right, *Dog spike, ca. 1921.* It was illustrated in Simmons-Boardman's 1921 Maintenance of Way Cyclopedia.

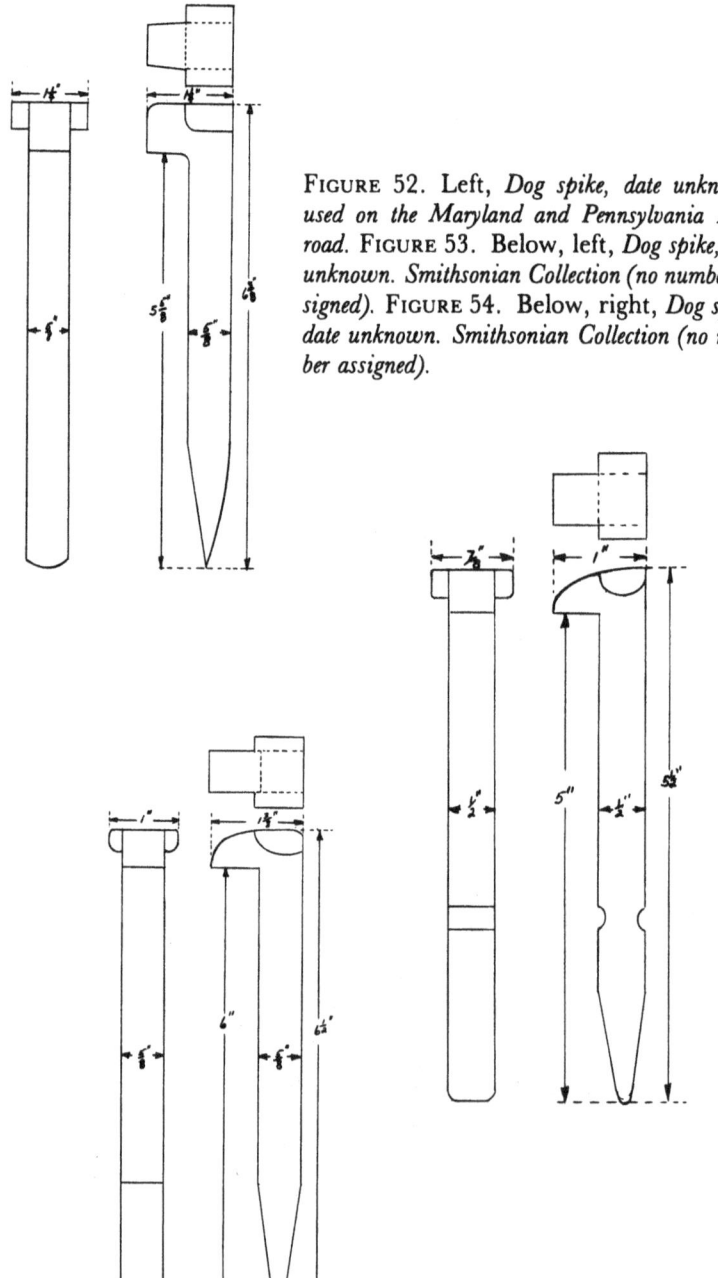

FIGURE 52. Left, *Dog spike, date unknown, used on the Maryland and Pennsylvania Railroad.* FIGURE 53. Below, left, *Dog spike, date unknown. Smithsonian Collection (no number assigned).* FIGURE 54. Below, right, *Dog spike, date unknown. Smithsonian Collection (no number assigned).*

## Greer Spike

The unusual swelling shape of the Greer spike (fig. 55) is claimed to increase the grip of the fibers, while the extra width gives greater resistance to lateral thrust.[33] At least that is what Howard Greer intended when he applied for and received patent No. 387,066 in 1888 for his unique design.

The spike has been used by the Kansas City, Fort Scott and Memphis Railroad; the Ophir Loop-Lizard Head pass portion of the Rio Grande Southern Railroad (RGS); and the Silverton Railroad.

FIGURE 55. *Greer Railroad Spike Company 1896 advertisement.*

An interesting story goes along with the use of the spike by these railroads. As the story goes, Mr. E. T. Jeffery was the president of the Rio Grande Southern at the time the spikes were purchased for the railroad. As he soon learned, the spike was easily bent by the

---

33. Tratman, *Railway Track and Track Work* (1908), p. 90.

outward thrust of the rail base. After this discovery, general usage of the spike quickly dwindled. However, Mr. Jeffery's name became associated with the Greer spike in the West, where it is commonly referred to as the Jeffery spike.

## *Daly Spike*

One way of increasing the holding power of a spike is to increase the shank surface area so that more wood fibers grip the spike. The Frost Railway Supply Company thought in 1905 that it had finally come up with the ultimate design in the Daly spike.[34] The unique spike (fig. 56) had a deep groove or channel running the entire length of the back of the shank giving six bearing faces in the wood instead of four. A series of tests by the Forest Service in 1906 measured the holding power of the spike and found that the channeled spike had about twelve percent more holding power than the common spike.[35] Even though the spike had superior holding power, it apparently never caught on as a practical design for general use.

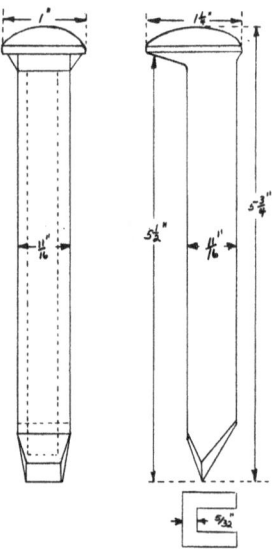

FIGURE 56. *Daly spike, ca. 1915.*

The Daly spike was most likely named for David A. Daly of Detroit, Michigan, who received patent No. 634,522 in 1899 for a sheet metal railroad spike which claimed a channeled design.

## *Bayonet Spike*

"The object of this invention is to furnish spikes so constructed that they will not split the ties into which they are driven, will not make a larger hole than they fill, can be readily and accurately

---

34. "The Daly Railroad Spike," *Railway Engineering and Maintenance of Way,* October, 1905, pp. 145-146.
35. W. Kendrick Hatt, "Holding Force of Railroad Spikes in Wooden Ties," *U.S. Department of Agriculture Forest Service—Circular 46* (Washington, D.C., 1906), p. 4.

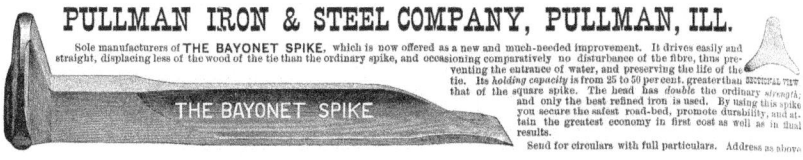

FIGURE 57. *Bayonet spike advertisement of 1885.*

driven, will hold securely, and will be economical of stock in their manufacture." Those were the design objectives of James Perkins for a new type of spike for which he received patent No. 236,511 in 1881. His first patent was followed by a second in 1884 (Patent No. 303,663) which claimed improvement over the first by changing the sharpened end and location of the cutting edge. An advertisement in the November 7, 1885, *Railway Review* (fig. 57) shows a spike which resembles closely the Perkins patent.

The Smithsonian Institution has one spike (Collection No. 326,640) identified as a Bayonet spike which had been used on the Grape Creek Branch of the Denver and Rio Grande Railroad, a three-foot narrow gauge railroad that ran from Cañon City to Westcliffe, Colorado.

## Faries Spike

Very few patented spikes ever made it into general use. A notable exception is the Faries spike originally patented (Patent No. 1,899,264) in 1933 by Robert Faries of St. Davids, Pennsylvania (fig. 58).

The Faries spike is made with a heavier throat, a gradual enlargement of the shank, and two angular lugs beneath the sides of the head. When driven, the full-throat design causes the spike to completely fill the spike hole in the tie plate without the spike head

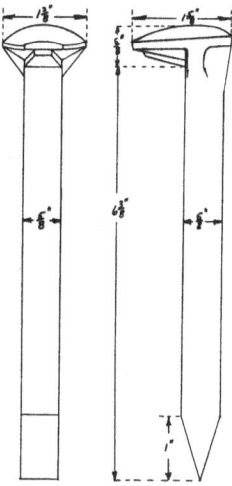

FIGURE 58. *Faries spike, ca. 1933, made by the Bethlehem Steel Corporation and used on the Pennsylvania Railroad.*

bearing on the rail flange, which permits the rail to flex under traffic without disturbing the spikes.

The principle advantages claimed for this spike are more metal at the point of greatest wear; greater strength of the head for pulling purposes; and not being backed out by rail flexure, thus retaining full holding power in the tie.[36]

The spike was still being made by the Bethlehem Steel Corporation in 1977.

## AREA Common Spikes

Three quarters of a century of railroading progress brought a multitude of construction methods, equipment designs, and operating devices. Among top railroad officials it became evident that the time had come for an association dedicated to the "advancement of knowledge pertaining to the scientific and economical location, construction, operation, and maintenance of railroads."[37] At the urging of *The Railway Age* magazine, a preliminary meeting was held in Chicago, October 21, 1898, with the purpose of organizing a railway association with these goals.

About twenty officials of various railroads came to the meeting and upwards of one hundred sent communications approving the movement and expressing a desire to be identified with the organization. The association was formed with the title "The American Railway Engineering and Maintenance of Way Association" which was eventually shortened to "American Railway Engineering Association" (AREA).

Standing committees were established with responsibility of investigating and reporting on specific subjects. Railroad spikes came under the responsibility of the Track Committee, and work on developing general specifications for a common spike began in 1911. After considerable controversy, the first design was finally approved in 1918. Improvements in the design have been made over the years with the 1937 design remaining as today's standard.

The development of AREA spikes is shown in figures 59–62. The Bethlehem Steel Corporation added an unusual twist to the AREA design to eliminate tie plate shifting. (fig. 63).

---

36. "Tie Plates and Fastenings," *Railway Engineering and Maintenance Cyclopedia* (Chicago, 1948), p. 322.
37. AREA, *Bulletin 480* (Chicago, 1949), p. 45.

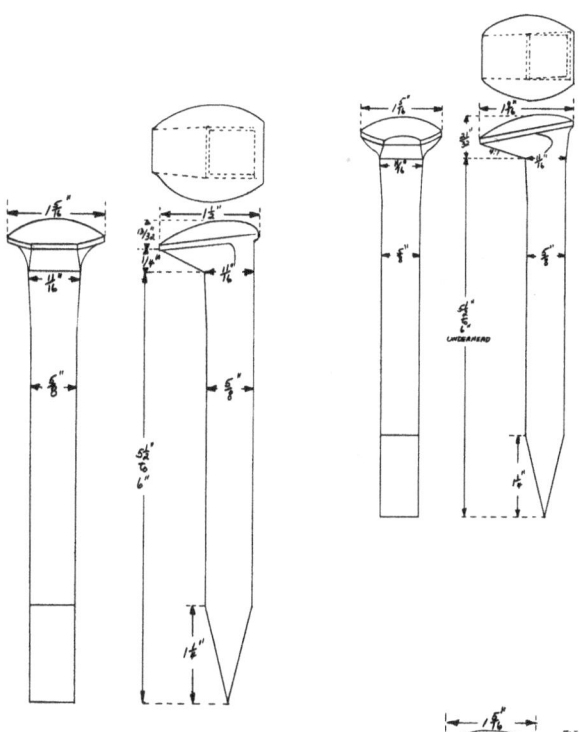

FIGURE 59. Above, left, *AREA ⅝ inch spike, 1918.* First AREA standard design, adopted in 1918. FIGURE 60. Above, right, *AREA ⅝ inch spike, 1921.* AREA 1918 design was rescinded and this 1921 design adopted because it could be more easily manufactured on spike-making machinery. A 9/16 inch spike was also adopted. FIGURE 61. Right, *AREA reinforced throat ⅝ inch spike, 1937.* This design provided greater clearance for a claw bar. It replaced the 1921 design and is still the current standard. A 9/16 inch spike was also adopted.

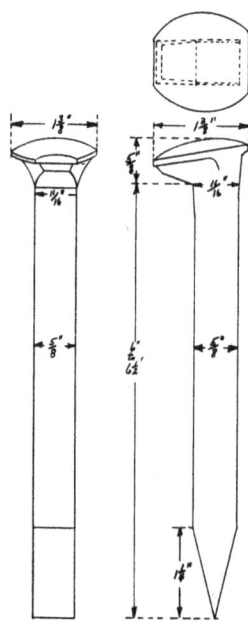

FIGURE 62. Left, *AREA reinforced throat spike, 1939, for toeless joint bars.* It was approved in 1939 but rescinded in 1962 because railroads were not using the design. FIGURE 63. Below, *AREA reinforced throat spike with a 15-degree torsional twist, 1977.* The design is supposed to eliminate the problem of the tie plate shifting while maintaining required track gage. It is manufactured by Bethlehem Steel Corporation.

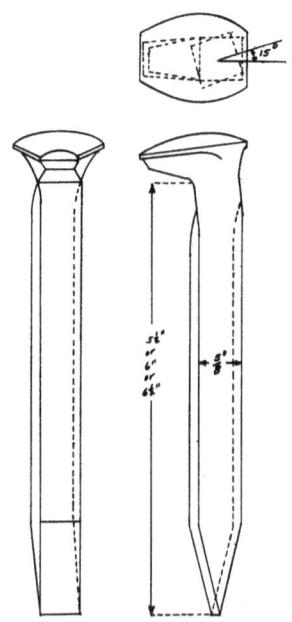

## Screw Spikes

The screw spike, first used on French Railways around 1860,[38] quickly became a standard in Britain, Europe, and other countries where the preservation of wood was important and manpower was cheap. In this country, the screw spike was first used on tests on a number of railroads (Kansas Pacific in 1870 and in 1890 on the Pennsylvania Railroad and the New York Central Railroad)[39] and then used in quantities on a few of them around 1910.[40] Even though the holding power of the screw spike is two to three times that of the common cut spike, a screw spike is not easy to install or remove. For this reason and others, it never gained wide popularity. Some screw spikes are shown in figures 64–71.

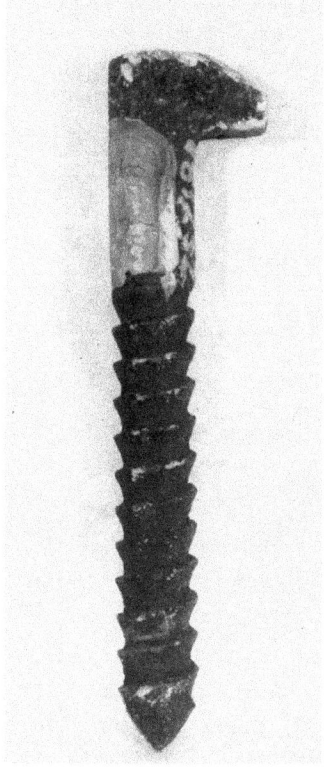

FIGURE 64. *Screw spike, ca. 1860, used on the South Carolina Railroad. Dimensions: 5¼ inches by ⅝ inches. Smithsonian Collection No. 181,232.*

---

38. William C. Willard, *Maintenance of Way and Structures* (New York, 1915), p. 182.
39. Tratman, *Railway Track and Track Work* (1908), p. 91.
40. "Track Section," *Maintenance of Way Cyclopedia* (Chicago, 1921), p. 164.

HENRY YOUNG, Pres't.  AGENCY OF THE  G. B. HUBBELL, Sec'y.
## U. S. RAIL-ROAD SCREW SPIKE COMPANY.
### No. 45 CLIFF STREET.
*Works at OSSINING, Westchester County, N. Y.*

To all persons engaged in the construction and management of Rail-Roads.

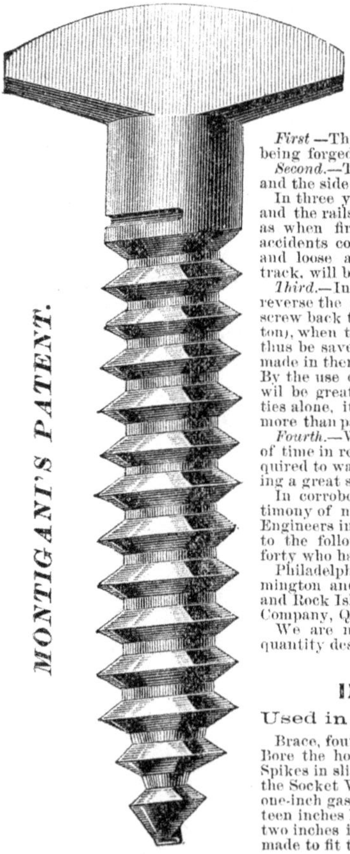

We respectfully call your attention to the double hook headed Screw Spike, for fastening rails and chairs to ties. The above cut represents the size and shape of the Screw. It has been thoroughly tested on several of the most important rail-roads in the country, and in every case it has stood the test in the most admirable manner.

The Screw Spike possesses the following advantages over the common spike:

*First*—They are made of the best refined Iron, the thread being forged (not cut), refines and toughens them.

*Second.*—The holding power of each screw is from 4 to 5 tons, and the side pressure 3 to 4 tons.

In three years' experience no screw has moved from its place, and the rails in every case have been held as firmly to the ties as when first put down. By its use, therefore, the numerous accidents constantly occurring from loose and broken spikes, and loose and broken rails, and consequent spreading of the track, will be obviated.

*Third.*—In order to remove the rails it will only be necessary to reverse the screw one-quarter, and when replacing it, turn the screw back to its original position (the same as turning a button), when the rail will be held firmly as before. The ties will thus be saved from the injury of splits, and the numerous holes made in them by repeatedly drawing out and replacing spikes. By the use of the screw, therefore, the durability of the ties wil be greatly increased. The saving to railway companies in ties alone, it is estimated (by experienced trackmasters), will more than pay the whole cost of the screw.

*Fourth.*—Where the screw is in use, in addition to the saving of time in relaying rails, a much less number of men are required to watch the track and keep it in repair, thereby causing a great saving.

In corroboration of the foregoing, we could present the testimony of many of the most experienced Track Masters and Engineers in the country, but deem it only necessary to refer to the following rail-road companies, selected from more than forty who have used these Spikes in more or less quantities, viz.:

Philadelphia and Reading, Hudson River, Philadelphia, Wilmington and Baltimore, New-York and New-Haven, Chicago and Rock Island, Fon Du Lac and Sheboygan, Albany Bridge Company, Quincy Bridge Company.

We are now prepared to manufacture these Screws in any quantity desired.

*MONTIGANI'S PATENT.*

### DESCRIPTION OF TOOLS
Used in Putting Down the Screw Spikes.

Brace, four feet long; Ship Carpenter's Bit, five-eighth inch. Bore the holes two and a half to three inches deep, start the Spikes in slightly with a hammer, then screw them home with the Socket Wrench, three feet long. The standard is made of one-inch gas pipe, the handle of three-quarter-inch gas pipe, fifteen inches long from centre; the foot is made of wrought iron, two inches in diameter, faced with steel, tempered; the slot is made to fit the head of the Screw Spike closely.

This Company manufactures and furnishes these tools at the following prices:

Long Brace,........................................$3 00 each,
Socket Wrench,...................................  3 00   "
Ship Carpenters' Bits,............................    70   "

In addition we furnish a Combination Wrench, excellent to use on bridges, price $20 each. For further particulars, address

### G. B. HUBBELL, Secretary,
No. 45 Cliff Street, New-York.

FIGURE 65. *Screw spike, ca. 1868. Only a quarter turn is needed to release the rail for replacement. A quarter turn back secures the new rail. A similar spike was used on the Kansas Pacific Railroad in the 1870s.*

FIGURE 66. Right, *Screw drive spike, ca. 1901,* driven the same way as a common cut spike. It was used on the Southern Pacific Railway. FIGURE 67. Below, left, *Screw spike, ca. 1906,* used on the Illinois Central. FIGURE 68. Below, right, *Screw spike, ca. 1915,* used on the Pittsburgh and Lake Erie with 1910 model tie plate.

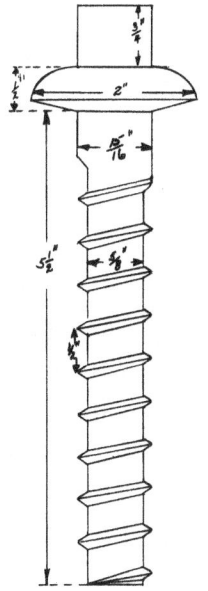

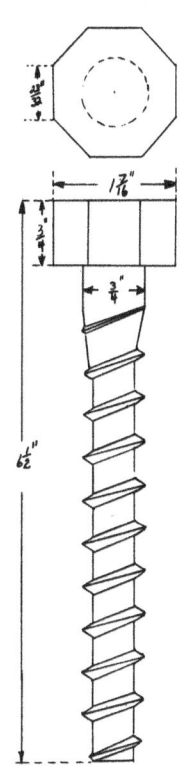

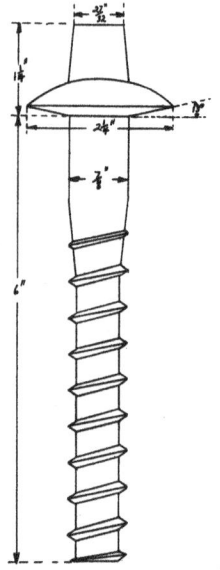

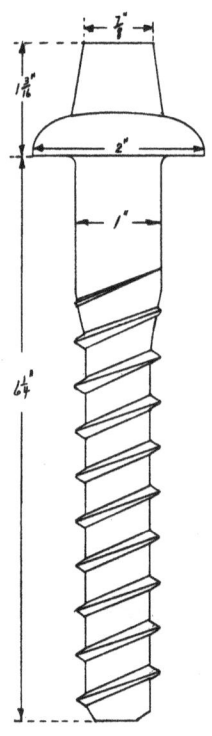

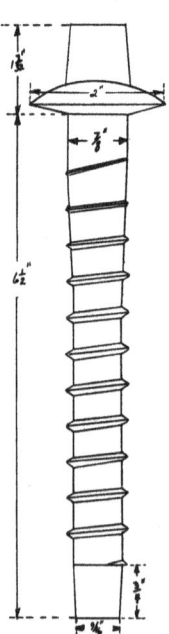

FIGURE 69. Above, left, *Screw spike, ca. 1915*, used on the Delaware, Lackawanna and Western Railroad. FIGURE 70. Above, right, *Screw spike, ca. 1919*, used on the Pennsylvania Railroad. FIGURE 71. Left, *Screw spike, 1977*, made by Bethlehem Steel Corporation.

## Sessler Spike

Instead of threads, the Sessler spike (fig. 72) had a grooved shank. The grooves had a twist so that the spike revolved slightly when driven. The spike had a round head not unlike that of a common screw spike and possessed many of the same characteristics.[41]

## Compression Spikes

Compression spikes came into general use in the 1930s with the claim of providing increased holding power when compared to the cut spike. The compression spike (fig. 73) is essentially one piece of heat-treated high carbon or spring steel.

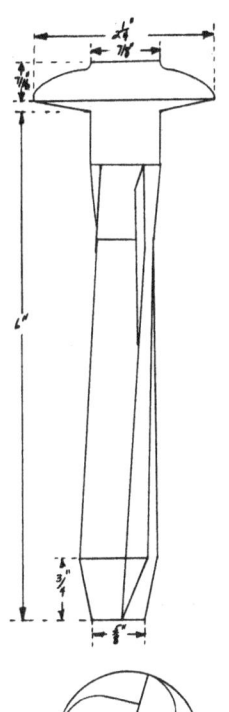

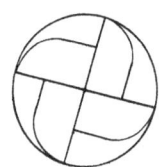

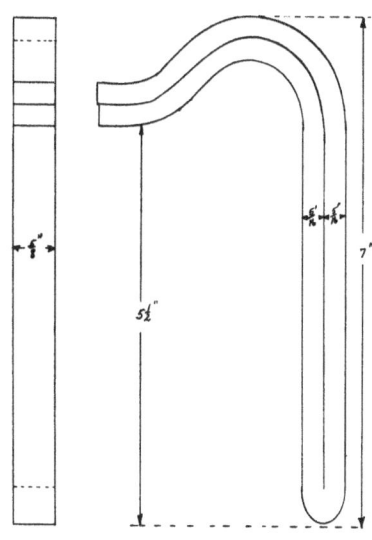

FIGURE 72. Above, *Sessler Grip spike, 1919, made by American Spike Company of New York City.* FIGURE 73. Left, *Compression spike, 1939, made by Elastic Rail Corporation.*

---

41. "Laboratory Tests on a New Spike," *Railway Maintenance Engineer*, December 1919, p. 443.

This type of spike is forged to form a straight shank ⅝ inch square with the upper portion or head of one or both leaves curved to press against the upper surface of the rail flange or tie plate. It is used only with prebored ties and can be driven with a regular spike maul or an air hammer adjusted to stop when the spike is properly seated. Special pullers have been devised for withdrawing these spikes without affecting their usefulness.

When driven to press upon the rail flange, the shank of the spikes remains solidly fixed in the wood while the curved head will flex with the movement of the rail under traffic. At the same time its spring action restrains rail movement vertically and longitudinally. When used as an anchor spike, the spring-type spike holds the plate firmly to the tie and has been found effective in minimizing mechanical wear of ties.[42]

## Gage Lock Spike

The gage lock spike [fig. 74] is a rail spike as well as a tie plate fastener. It is designed for use on tangent track and light curves where lateral thrust can be overcome with two spikes at each plate. These are driven in the line holes of the tie plate, adjacent to the base of the rail.

Gage lock spikes are installed in predrilled holes of 9/16 inch diameter . . . drilled to a minimum of 7 inches.

Two gage lock spikes hold the plate to the tie and maintain gage and line more effectively than four cut spikes per plate.[43]

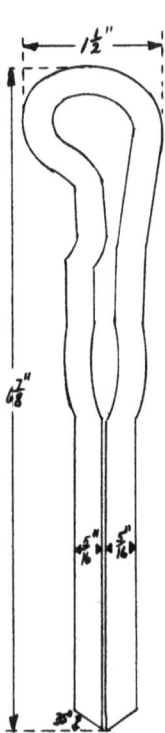

FIGURE 74. *Gage lock spike, 1977, made by Bethelehem Steel Corporation.*

---

42. "Tie Plates and Fastenings," *Railway Engineering and Maintenance Cyclopedia* (1948), p. 319.
43. Bethlehem Steel Corporation, *Railroad Fasteners Booklet No. 3193* (Bethlehem, Pa., 1977), p. 5.

# Bibliography

### BOOKS

American Railway Engineering Association. *Bulletin 480.* Chicago, March 1949.

American Railway Engineering Association. *Manual for Railway Engineering.* Chicago, 1978.

Ames, Charles E. *Pioneering the Union Pacific.* New York: Meredith Corp., 1969.

Ashcroft, John. *Railway Directory.* New York: (by the author), 1868.

Baltimore and Ohio Railroad. *Standard Plans for Maintenance of Way and Construction.* Baltimore, 1907. Reprint, Akron, Ohio: Akron Railroad Club, Inc., 1977.

Barry, John W. *Railways and Locomotives.* London: Longmans, Green and Co., 1882.

Camp, W. M. *Notes on Track Construction and Maintenance.* Chicago: (by the author, Auburn Park), 1903.

Cole, W. H. *Notes on Permanent-way Material, Plate-laying, and Points and Crossings.* New York: Spon and Chamberlain, 1896. The book was updated in 1928 and issued under the title *Permanent-Way Material, Plate-laying, and Points and Crossings.*

Couche, CH. *Permanent Way, Rolling Stock and Technical Working of Railways.* London: Dulau & Co., 1877.

Dodge, Grenville M. *How We Built the Union Pacific Railway.* Washington: Government Printing Office, 1910.

Hatt, W. Kendrick. "Holding Force of Railroad Spikes in Wooden Ties." *U.S. Department of Agriculture Forest Service Circular 46.* Washington: Government Printing Office, December 26, 1906.

Hay, William W. *Railroad Engineering.* New York: John Wiley & Sons, 1953.

Hubbard, Freeman H. *Railroad Avenue.* New York: McGraw-Hill Book Company, Inc., 1945.

Jervis, John B. *Railway Property.* New York: Phinney, Blakeman & Mason, 1861.

Kirkman, Marshall M. *Building and Repairing Railways.* New York: World Railway Publishing Co., 1903.

Low, James W. *Railway Directory.* New York: E. R. Bennet & Co., 1861.

Mills, William H. *Railway Construction.* New York: Longmans, Green and Co., 1898.

New York Central and Hudson River Railroad. *Standard Plans Book.* 1907.

Nicollis, William J. *The Railway Builder.* Philadelphia: Henry Carey Baird & Co., 1878.

Parsons, W. B. *Track.* New York: Engineering News Publishing Co., 1886.

Poor, Henry V. *History of the Railroads and Canals of USA.* New York: John A. Schultz & Co., 1860.

Proudfit, Margaret Burden. *Henry Burden, His Life and History of His Inventions.* Troy, N.Y., 1904.

Ringwalt, J. L. *Development of Transportation Systems in the United States.* Philadelphia: (by the author), 1888.

Road and Track Supply Association. *Complimentary Book.* Chicago: B. S. Wasson & Co., 1894.

Sellers, Morris, & Co. *The Track Spike, The Tie, The Splice Bar.* Chicago: (by the author), 1891.

Sellew, William H. *Railway Maintenance Engineering.* New York: D. Van Nostrand Co., 1919.

Smith, F. A. *Michigan Central Railway Track Department Maintenance of Way Standards.* Chicago: Roadmaster & Foreman, 1896.

Stevenson, David. *Sketch of the Civil Engineering of North America.* London, 1838.

Tanner, Henry S. *A Description of the Canals and Railroads of the United States.* New York, 1840.

Tratman, E. E. Russell. *Railway Track and Track Work.* New York: The Engineering News Publishing Co., 1901, and revised in 1908.

Tratman, E. E. Russell. *Railway Track and Maintenance.* New York: McGraw-Hill Book Co., 1926.

Van Auken, Kenneth L. *Practical Track Work.* Chicago: Railway Educational Press, Inc., 1916.

Von Schrenk, Hermann. "Cross-Tie Forms and Rail Fastenings, With Special Reference to Treated Timbers." *U.S. Department of Agriculture Bureau of Forestry Bulletin No. 50.* Washington: Government Printing Office, 1904.

Watkins, J. Elfreth. *The Camden and Amboy Railroad Origin and Early History.* Washington: Press of Gedney & Roberts, 1891.

Watkins, J. Elfreth. *The Development of the American Rail and Track.* Washington: Government Printing Office, 1891.

Webb, Walter. *Railroad Construction.* New York: John Wiley & Sons, 1900.

Webber, R. I. "Holding Power of Railroad Spikes." *University of Illinois Bulletin No. 18*, June 15, 1906.

Weiss, C. *Practical Railway Maintenance*. New York: McGraw-Hill Co., 1923.

Willard, William C. *Maintenance of Way and Structures*. New York: McGraw-Hill Book Co., 1915.

Wilson, Winter. *Elements of Railroad Track and Construction*. New York: John Wiley & Sons, 1915.

## ARTICLES

"Are There Possible Economies in the Use of Track Walkers?" *Railway Maintenance Engineer*, November, 1922.

Bowman, J. N. "Driving the Last Spike at Promontory 1869." *Utah Historical Quarterly*, Winter 1969.

Ketterson, F. A., Jr. "Golden Spike National Historic Site: Development of an Historical Reconstruction." *Utah Historical Quarterly*, Winter 1969.

"Laboratory Tests on a New Spike." *Railway Maintenance Engineer*, December, 1919.

Lampson, Robin. "The Golden Spike Is Missing." *The Pacific Historian*, Winter 1970.

Morrison, A. "Wood Screws and Spikes for Railroad Track Use." *Railroad Gazette*, October 15, 1897.

"Pennsylvania Railroad Standard Spike." *Railroad Gazette*, December 4, 1885.

Rezneck, Samuel. "Office Building 1881 Burden Iron Company, Troy." *A Report of the Mohawk-Hudson Area Survey*. Ed. by Robert M. Vogel. Washington: Smithsonian Institution Press, 1963.

"The Daly Railroad Spike." *Railway Engineering and Maintenance of Way*, October 1905.

"The 'Churchill' Rail Joint and 'Diamond' Spike." *The Street Railway Journal*, Vol. XIV, No. 12, 1898.

"The Railroad Spike—An Anachronism of the Twentieth Century." *Scientific American*, August 8, 1908.

## CATALOGS

Bethlehem Steel Corporation. *Railroad Fasteners Booklet No. 3193*. Bethlehem, Pa., June 1977.

Carnegie Steel Company. *Pocket Companion for Engineers, Architects and Builders*. Pittsburgh, Pa., 1913.

Gregg Company. *General Catalogue No. 550*. Hackensack, N.J., ca. 1920.

Illinois Steel Company. *Track Materials Catalogue*, 1919.

PERIODICALS

*American Railroad Journal.* New York, 1832-1886.

American Railway Engineering Association. *Proceedings of the Annual Conventions*, Chicago, 1900 to present.

*Maintenance of Way Cyclopedia* renamed *Railway Engineering and Maintenance Cyclopedia.* Chicago, 1921-1955.

*Pocket List of Railroad Officials.* New York, 1895 to the present.

*Poor's Manual of Railroads of the United States.* New York, 1871-1897.

*Railroad Gazette.* New York, 1870-1908.

*Railway Age.* Chicago and New York, 1876 to the present.

*Railway Maintenance Engineer.* Chicago, 1905 to the present. Now called *Railway Track and Structures.*

*Railway Review.* Chicago, 1868-1926.

*Scientific American.* New York, 1845 to the present.

www.ingramcontent.com/pod-product-compliance
Lightning Source LLC
Chambersburg PA
CBHW051716040426
42446CB00008B/911